HUMAN BIOLOGY

Deryck Taverner, MBE, MD, FRCP, formerly Senior Physician at Leeds General Infirmary and Reader in Medicine at the University of Leeds, is the author of several established medical and nursing textbooks, including *Taverner's Physiology* (Hodder & Stoughton).

TEACH YOURSELF BOOKS

HUMAN BIOLOGY

Deryck Taverner

TEACH YOURSELF BOOKS
Hodder and Stoughton

First edition 1964
Second edition 1974
Third edition 1984

British Library Cataloguing in Publication Data
Taverner, Deryck
Human biology. – (Teach yourself books) 3rd ed.
1. Human physiology
I. Title
612 QP34.5

ISBN 0 340 32977 7

Printed and bound in Great Britain for
Hodder and Stoughton Educational,
a division of Hodder and Stoughton Ltd,
Mill Road, Dunton Green, Sevenoaks, Kent,
by Hazell, Watson & Viney Ltd,
Aylesbury, Bucks. Photoset by
Rowland Phototypesetting Ltd,
Bury St Edmunds, Suffolk.

Contents

List of Illustrations

Preface to the Third Edition

This third edition has been extensively revised and re-illustrated, and much new material has been added. It now covers the syllabus for the Ordinary Level and Alternative Ordinary Level GCE examinations. The book is still intended also for the ordinary reader without special scientific knowledge, and emphasis has been placed upon clarity and simplicity as far as is possible in this rapidly expanding field. The treatment is meant as a guide and a stimulus to further study. Any impression of dogmatism reflects the need for clarity rather than a feeling of certainty.

D. Taverner

1

The Human Body

The human body is an extremely elaborate, sensitive, efficient, self-regulating mechanism which is continually under attack by adverse forces. It maintains its internal climate and organisation constant within very narrow limits in spite of severe changes in the conditions outside. Only the most severe external stress will destroy or seriously damage the normal human body. This capacity for self-regulation or *homeostasis* is one of the essential properties of living organisms. Others are a cellular organisation, ingestion, excretion, respiration, locomotion, response to stimuli, reproduction and growth.

Cellular organisation

The basic unit of plant and animal life is a minute speck of living jelly – a *cell* (p. 19). A single cell is vulnerable to slight changes in its environment, but in the course of billions of years of evolutionary change more and more elaborate associations and groupings of cells have developed. The adult human body consists of more than one trillion cells organised and coordinated to form a functional whole. There are many different kinds of cells, each with its own particular structure and function. The cells are grouped and arranged to form *tissues* and *organs* to facilitate performance.

The differentiation of structure and function builds up complex organisms in which specialised cells, tissues and organs are dependent on other cells, tissues and organs to maintain their basic working conditions. The movements of foodstuffs, gases, fluids, dissolved molecules and information about the body, plus the need for close control of temperature and chemical conditions, require elaborate circulatory, respiratory and nervous systems. Rapid, accurate communication is essential for a speedy, effective reaction to, or elimination of, potentially dangerous factors in the environment. The more elaborate, precise and interlocked these various systems become, the greater are the possibilities of derangement and disease.

Homeostasis

Nearly all the cells in the body are insulated and protected to a high degree. Sheets of dead cells cover the surface of the body and the few areas where living cells are exposed are protected by special coverings of fluid and mucus. The rest are bathed in tissue fluid which is maintained within very narrow limits of temperature, chemical composition and physical conditions. Variations in the composition of tissue fluid resulting from changes in the internal or external environment are rapidly corrected by various mechanisms and countermeasures to be described.

To achieve this the body is equipped with devices which respond to changes from the normal state and send signals to other parts to evoke the necessary reaction. Much of this signalling is done by the *nervous system*, which transmits information and instructions throughout the body with speed and accuracy. Other homeostatic mechanisms are *hormones* – chemical compounds which travel more slowly in the blood stream (Chapter 15).

Ingestion

Living beings extract energy from their surroundings and use it for their own needs of maintenance and survival. We ingest foodstuffs of plant and animal origin. Foodstuffs consist mainly of atoms of carbon (C), hydrogen (H), oxygen (O), nitrogen (N) and smaller quantities of other elements (p. 7) which are bound together into larger molecules. The chemical bonds of these molecules form the pool of energy utilised by the body. All the energy of living organisms comes ultimately from the sun. In the presence of sunlight plants are able to make sugar – glucose – from water and the carbon dioxide in the air by the process of *photosynthesis* (Fig. 1.1). The green pigment – chlorophyll – in the chloroplasts of the leaves, utilising light and warmth from the sun, provides the energy

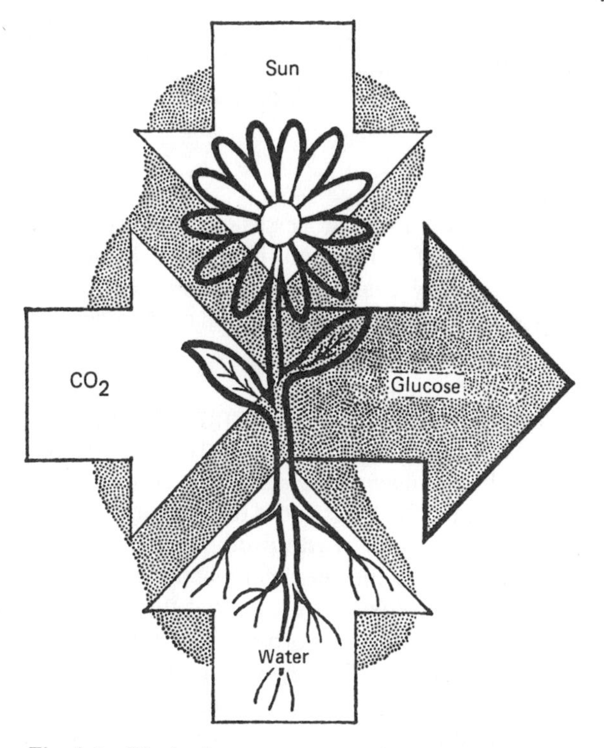

Fig. 1.1 The basic energy pattern (photosynthesis)

for the formation of glucose with the release of oxygen into the air. The glucose passes from the leaves of the plant to its stem and roots, where it is converted to starch for storage. Plants also extract nitrogen and minerals from the soil to form *amino acids* and *proteins*. Additionally they can make fats and oils.

In animals, foodstuffs and water are generally consumed in amounts that are related to the needs of the body. Foodstuffs enter a long coiled tube in which serial chemical reactions break them down into simpler forms which can be absorbed into the body. They are then carried by the blood stream through a closed, branching system of tubes – *arteries* and *veins* – to all the tissues and organs of the body. Here, further elaborate chemical reactions break them down to provide energy, or transform them into more complex forms which can be stored or used as building materials. The blood is moved through the blood vessels by the regular beating of a muscular pump, the *heart*. Its rate and output are continuously and accurately adjusted to the needs of the body by an intricate system of controls.

Excretion

The end products of these processes include waste materials which have to be removed from the body if it is not to be poisoned by them. The main excretory channel is the kidney (p. 92), but breath, sweat (pp. 65, 176) and faeces (p. 75) all contain waste products.

Respiration

The chemical reactions releasing energy normally require a steady supply of oxygen to all parts of the body. The oxygen is extracted from the air by the *haemoglobin* in the blood passing through the two spongy organs, the lungs and is then carried by the blood to the tissues where it is used. Waste products and carbon dioxide are continually formed by the chemical reactions in the tissues. Large amounts of carbon dioxide are excreted through the lungs at the same time as oxygen is absorbed. The physical and chemical processes involved are delicately coordinated so that they continue precisely and efficiently for long periods by means of their self-compensating adjustments. The details are described in Chapter 10.

Response to stimuli

The existence and survival of each individual depends upon its capacity to respond effectively to outside changes or stimuli. This excitability, or irritability, is a fundamental property of living matter. Individual cells, tissues, organs and the organism as a whole can all respond to stimuli. The response depends upon the capacity of the recipient, but at the highest level all activity such as movement, speech, vision and consciousness result from a constant stream of stimuli reaching the central nervous system from all parts of the body.

Reproduction

Although not essential for the individual, the capacity for reproduction is essential for survival of the species. Reproduction by the intermingling of the genetic material from two individuals is the fundamental source of biological variation and the basis of evolution. The mixture of precision and chance involved is described on page 26. The fusion of material from the two bodies blends some of the physical characteristics, defects and capabilities of both parents into a new individual with a unique structure and the possibility of developing its inherent capacities as far as its surroundings and environment permit. This interplay of the individual, with a unique genetically determined biochemical endowment, and its unpredictable, varying environment is the key to the development and function of the human being.

Locomotion

All life has the power of motion to some degree and animals possess the capacity for independent movement. In humans this results from the coordinated, graded contraction of muscle under the control of the central nervous system. The senses of smell, vision and hearing are linked to information derived from the balancing organs and from *receptors* throughout the skin, muscles, bones and joints to provide a basis for independent, purposive movement and performance.

Growth

Every mature individual has grown from a single fertilised cell or ovum by repeated division. Some of the processes concerned are described in Chapter 18.

Interdependence of animals and plants

Most animals are herbivorous; that is, they feed upon plants. Some live on the flesh of other animals but the majority eat grass, leaves, nuts, seeds, fruit or the minute green plant cells in the sea. The basic source of virtually all food is the conversion by plants of the energy of sunlight into the chemical energy of organic compounds – photosynthesis. Plants take up carbon dioxide from the air and combine it with water in the presence of sunlight to form plant tissue, giving off oxygen. The oxygen derived from plants is essential to animals to maintain their energy-producing chemical processes.

The chlorophyll of plants acts as a catalyst, with a series of *enzymes* in the chloroplasts splitting water molecules into hydrogen and oxygen. Some of the hydrogen atoms combine with carbon dioxide to form glucose and much of the oxygen is released into the atmosphere. The residual hydrogen and oxygen atoms reform water and release the energy taken up from sunlight to form high energy phosphate compounds such as adenosine triphosphate (ATP). This energy is then used to form more complex organic molecules from glucose and nitrates, phosphate and minerals taken from the soil.

The waste products from animals, and eventually their carcasses, are broken down by bacteria and fungi so that the basic chemicals are finally available for return to the plant-animal cycle. Bacteria colonise the animal body in mutual cooperation or *symbiosis*, and do not normally cause disease. Most bacteria flourish harmlessly, as in the intestine, where they are valuable agents in the processes of digestion. Other types of organisms are always harmful and are parasites. They can cause serious health problems and are discussed in Chapter 17.

2

Body Chemistry

Elements

The simplest permanent unit of any material is an *atom*. Usually two or more atoms are linked to form a *molecule*. An *element* is composed of molecules of the same atoms. Elements are often found mixed together, as in air, which is a mixture of oxygen and nitrogen. In such mixtures the atoms are continually moving about so that they eventually become thoroughly mixed together. Each kind of atom is designated by its own capital letter. The elements found in the human body are carbon (C), hydrogen (H), oxygen (O), nitrogen (N), sodium (Na), potassium (K), chlorine (Cl), phosphorus (P), sulphur (S), iron (Fe) and iodine (I), plus traces of some others.

Most chemical substances are *compounds* formed of numbers of atoms linked together as molecules. Energy is needed to form such linkages – chemical bonds – and the energy used by the body comes from their breakdown. Molecules may be very simple, consisting of two or three atoms, or very complex, containing millions of atoms. The composition of molecules can be represented by chemical formulae which indicate the numbers of each atom forming them. Examples of simple molecules are sodium chloride (common salt), $NaCl$; water, H_2O; and carbon dioxide, CO_2. The sugar glucose is a more complex molecule containing six carbon, twelve hydrogen and six oxygen atoms – $C_6H_{12}O_6$. Molecules may themselves be joined together to form very large molecules. The starch of plants is made up of many glucose molecules linked together. A compound related to starch, called *glycogen*, found in the liver and muscles, is also built up of glucose molecules and forms a reserve supply of energy.

All plants and animals are made up of *water* and three kinds of chemical substances: *carbohydrates*, *fats* and *protein*. Three-quarters of the animal body is water and its chemical reactions occur in watery solution within very narrow limits of temperature and acidity or alkalinity. These ceaseless reactions take place in a sticky jelly called *protoplasm*, which is the elementary material of all living matter. It is arranged in separate units – cells – that vary greatly in size. The cells are grouped together in great numbers as *tissues* – fat, gristle (cartilage), muscle and bone (Fig. 2.1). The tissues are arranged in organs, each with a special function.

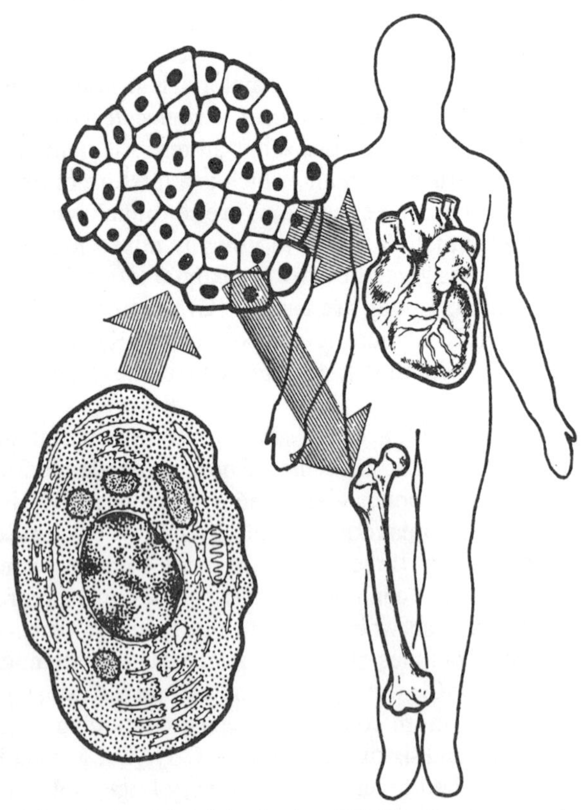

Fig. 2.1 The relationship of the single cell (*left*) to the tissues and organs of the body

Enzymes

The chemical reactions responsible for the immensely varied phenomena of life depend upon the presence of specific proteins called enzymes. Many of these reactions cannot be performed in the laboratory or only under carefully controlled conditions of temperature and acidity. Starch is very stable and can only be broken down by prolonged boiling with strong acid; yet in the mouth starch is rapidly decomposed by a single enzyme – ptyalin or salivary amylase. Enzymes act as catalysts, speeding up chemical reactions but taking no direct part in them. Many of the chemical reactions in the body involve numerous enzymes, each one responsible for a single tiny step. There may be dozens of simultaneous chemical reactions going on in an active cell and each step in this intricate performance depends upon the inter-related, self-regulating activity of the enzyme systems of the cell. Some enzymes act outside the cell in which they are made, as do those involved in the digestion of foodstuffs. Some are so powerful that they would destroy their parent cell if they were not stored in an inactive form. They are converted into the active form, after leaving the cell, by special activators – *kinases*. For example, the inactive trypsinogen of pancreatic juice (p. 72) is converted into trypsin, which digests protein, by the enterokinase of intestinal juice.

There are many thousands of different enzymes in the human body, each with a specific function in relation to a single reaction or process. Some build up larger molecules from small ones – *synthesis* – and others break the large into smaller molecules – *degradation* – often with the release of energy. All the varied chemical activity of the body involved in these processes is called the *metabolic* activity. Breaking down is referred to as *catabolism* and *anabolism* means building up.

Energy

The chemical and physical processes of the body require a regular supply of energy and without energy the cells of the body soon die. Energy is essential to preserve the unstable organisation of the cell and to perform mechanical, electrical and chemical work.

Ultimately all our energy derives from the action of sunlight on plants producing glucose and more complex molecules. These may be assimilated directly from plants or via the flesh of animals, which use ready-made fuel from the cells and tissues of plants and other animals. Their cells can extract energy from the chemical bonds built into these fuels when the molecules are synthesised.

Liberation of energy

The digestion of food produces simpler molecules, which are absorbed and eventually oxidised (i.e. burnt with oxygen) to produce carbon dioxide and water. The large molecules of the foodstuffs are broken down along specific pathways which are controlled by series of enzymes. The number of chemical changes involved is, in fact, less than might be expected because many of the metabolic

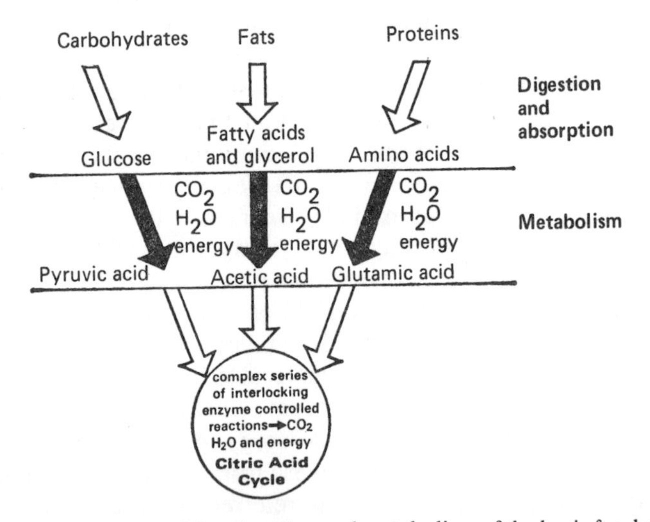

Fig. 2.2 The stages of the digestion and metabolism of the basic foodstuffs to provide energy

pathways converge. Fig. 2.2 shows how the initial processes of digestion and absorption converge into a series of reactions common to all three types of foodstuffs. This is called the *citric acid cycle* after one of the key chemical compounds in the series. There are intermediate stages involved, each controlled by a specific enzyme, with the formation of carbon dioxide and water at each stage. The metabolism of proteins gives rise to more complex waste products, especially urea, in addition to carbon dioxide and water.

Whenever a chemical bond is broken, energy is released. Two-thirds of this energy is dissipated as heat, but one-third is taken up by a chemical compound called adenosine triphosphate (ATP). This consists of an amino acid linked by chemical bonds to three phosphate molecules. Very large amounts of energy, derived ultimately from the foodstuffs in the diet, can be stored in this compound. Whenever energy is required, a phosphate molecule is split off by enzyme action and ATP is converted to adenosine diphosphate (ADP); energy is then released.

When energy is extracted from cellular fuel by the activity of enzymes, it is stored by means of the conversion of ADP back to ATP. The actual source of the energy is the chemical force linking or bonding the phosphate group to the ATP molecule. The energy-rich ATP store of each cell is kept fully charged; the glucose and other foodstuffs entering the cell can be metabolised only if ADP molecules are ready to accept the energy they release. Otherwise the glucose is stored as glycogen and the other substances accumulate.

Diffusion

All matter consists of minute particles (atoms or molecules) in constant motion. In solids the particles vibrate rapidly but do not move from their fixed positions; in fluids they move more freely and in gases very freely indeed. Heat increases the intensity of these vibrations and the increased molecular movement on heating causes, for example, steam to rise from boiling water. This movement, or energy, ensures that in time all the particles in a given liquid, gas or mixture of gases, will be distributed evenly throughout the whole volume of the material. This process is called diffusion and it is responsible for the movement of substances in solution. When glucose is added to water a solution is formed in which the glucose will spread until the distribution of its molecules is uniform. Diffusion in liquids is slow but is effective over short distances. In gases diffusion is much quicker because there is much more space between the gas molecules.

Osmotic pressure

When a soluble substance such as salt or sugar is added to water it diffuses slowly throughout the water and achieves equal concentration everywhere in it. If barriers such as cell membranes interfere with diffusion considerable osmotic pressure may be set up. Thin sheets of membrane from various sources, some of them animal, are semipermeable. They have innumerable tiny holes through which water and small molecules pass freely but which hold back larger molecules such as glucose. The widely used packaging material cellophane is a good example of a semipermeable membrane. Some of these membranes will allow only oxygen, carbon dioxide and water to pass; some will allow through water and smaller molecules such as sodium, urea and glucose, but not larger particles such as protein molecules.

If a semipermeable membrane through which only the water molecules can pass freely is placed between water and a solution of, say, sugar in water, then water molecules will pass readily from the water into the sugar solution. The tiny holes in the membrane are continually bombarded by molecules of all sizes. The large sugar molecules cannot pass through the holes on their side and, in effect, fewer holes are available for the smaller water molecules to pass from left to right (Fig. 2.3) than from right to left. If the membrane divides some kind of container into two halves, water will accumulate in greater amount on the side containing a solution of sugar. The build up of fluid on this side sets up an 'osmotic' pressure.

The osmotic pressure of a solution depends upon the number and size of the molecules in the solution, e.g. sodium chloride in a salt solution of a given strength is in the form of a larger number of tiny particles (ions) which exert a stronger osmotic pressure than would a solution of albumin (p. 81) of the same concentration but containing a smaller number of larger molecules. Similarly, globulin (p. 81), which has a much larger molecule than albumin, would exert a correspondingly lower osmotic pressure than would a solution of albumin of equal strength. In the body the cell membranes are semipermeable and the flow of water by osmosis is of great importance in various activities of the body. The cell membranes are freely permeable to water and to some ions, but not to proteins.

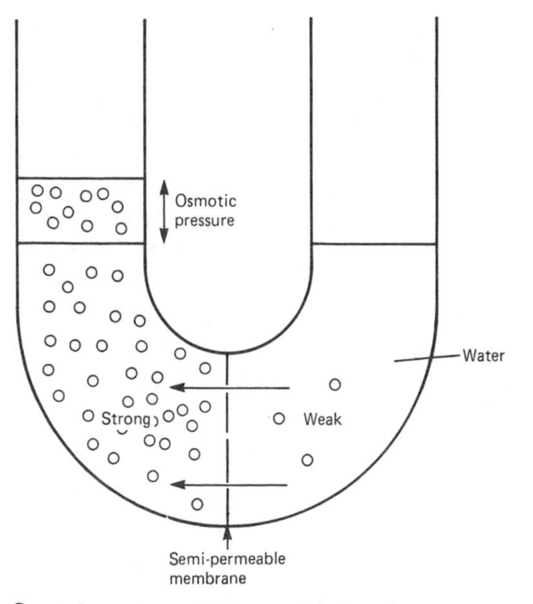

Fig. 2.3 Osmotic pressure. Water molecules can move freely across the membrane. The white circles represent the larger sugar molecules, which cannot pass through the membrane but constantly collide with it and rebound into the solution of sugar. As a result more water molecules pass through the membrane from right to left causing a net accumulation of water molecules on the left-hand side. This causes a flow of water across the membrane and increases the volume of the sugar solution. In effect, in any given unit of time, more holes are open from right to left than in the opposite direction. The difference in the levels of the fluid in the two columns represents the osmotic pressure attributable to the sugar in the solution.

In effect there is a tendency for water to move in such a way that adjacent solutions achieve the same concentration. When a cell is placed in a solution, water will pass in or out of it according to the osmotic pressures of the cell contents and the surrounding medium. Solutions exerting the same osmotic pressure as the cell contents permit water to pass equally in both directions so that the volume of the cell remains the same; such solutions are said to be *isotonic* with the cells. Solutions which exert higher osmotic pressure make more water leave the cell than enters and the cell shrinks; such solutions are termed *hypertonic*. Solutions which have a lower osmotic pressure allow more water to enter than leave the cell so that it finally bursts; such solutions are said to be *hypotonic*.

It is found that the osmotic pressure of blood plasma is equivalent to the osmotic pressure exerted by a solution of 0.9 per cent sodium chloride (NaCl), and this strength is taken as the isotonic level in physiological calculations. The red blood corpuscles, which are surrounded by a semipermeable membrane, are thus in balance with an 0.9 per cent solution of sodium chloride. If they are placed in stronger (hypertonic) solutions they will lose some of their contained water to the stronger solution and become wrinkled up, or crenated. In solutions weaker than 0.9 per cent (hypotonic) red blood capsules swell up and burst. This is known as *haemolysis*.

Transport through the cell membrane

The life and function of each cell depend upon the adequate transport of oxygen, nutrients, ions and waste products through its outer membrane, which consists of a double layer of phospholipid molecules arranged side by side in a regular pattern. Protein globules are embedded in these layers, sometimes extending from one side of the membrane to the other. They may move freely within the membrane and perform a variety of functions. Some of these proteins are structural, connecting cells together and helping to maintain the form of organs. Many are receptors for hormones and signal transmitters or act as enzymes to facilitate chemical reactions at the cell surface. Others form selectively permeable channels for the passage of ions. These proteins can change shape when active and open or close the channel to a particular ion. There are also continually acting pumps which maintain differential concentrations of ions across membranes. Some proteins act as surface markers concerned with defence mechanisms.

Water and small molecules such as sodium, chloride and urea pass freely in and out by diffusion but larger molecules such as glucose, fat and protein cannot penetrate. Fatty substances are dissolved by the lipids in the cell membrane, pass through and are released inside the cell. Other substances pass through by 'active transport'. They combine with a 'carrier' to form a compound which dissolves in the cell membrane and is passed through leaving the carrier behind (Fig. 2.4). Glucose, for example, combines with phosphate, under the influence of an enzyme, to form glucose-phosphate, which is

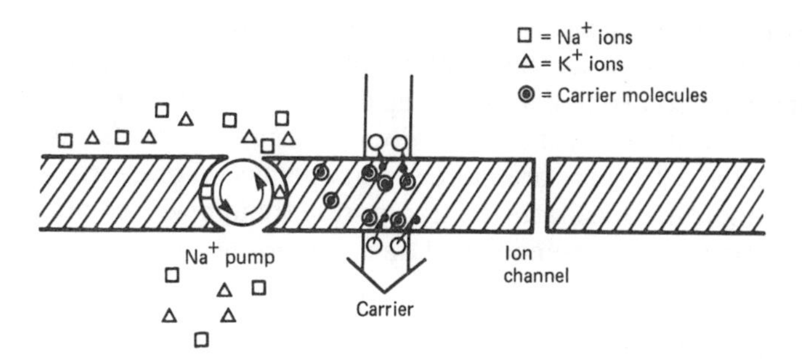

Fig. 2.4 Transport through a cell membrane. Black dots represent carrier molecules

soluble in the cell membrane. At the inner surface of the cell membrane the reverse process takes place and glucose is liberated into the cell.

Membrane potential

Most cells show a difference in electrical potential (polarisation) across their membrane with the inside 70–90 mV negative to the outside. This level of membrane potential is found in excitable cells such as nerve and muscle but in non-excitable cells such as erythrocytes and kidney the value is only −10 to −30 mV. These potentials arise because of differences in distribution of ions across the cell membranes resulting from differences in membrane permeability. During activity in excitable cells depolarisation occurs because of a sudden inflow of Na^+ ions and the resting potential alters but recovers again rapidly by repolarisation (p. 136), when the ionic distributions are restored.

Sodium pump

The purely physical processes of diffusion through the protein channels are supplemented by large numbers of continually acting pumps. There is normally ten times more sodium outside a cell than inside and ten times more potassium inside than outside. These levels are maintained by protein molecules which, using the energy of the phosphate bond of ATP molecules, exchange three Na^+ ions inside a cell for two K^+ ions outside. Cells may have up to 200 of these 'sodium pumps' per square micrometre of surface (μm^2) and an excitable cell like a neurone may have one million pumps operating up to 30 times per second. They can move more than 200 million N^+ ions per second out of a cell and up to 130 million K^+ ions in. This is the basis of the electrical activity of excitable cells. In layers of epithelial cells there are more sodium pumps on one surface than the other; this facilitates the transport of sodium with chloride and water in the intestine and renal tract.

Acid-base balance

Acids are substances that turn blue litmus paper red; bases (alkalis) turn red litmus paper blue. Bases counteract or neutralise acids. In watery solutions all acids and bases split in varying amounts into electrically charged particles called ions. An acid is a substance that produces hydrogen ions (H^+) and a base is a substance that accepts hydrogen ions. Strong acids contain large quantities of hydrogen ions and weak acids contain small quantities. The chemical activity of the cells of the body constantly produces acids, which tend to increase the concentration of hydrogen ions (H^+) in the tissue fluids and blood. Even slight alterations of hydrogen ion concentration from the normal level produce striking changes in the rates of chemical reaction in the cells.

A special system of nomenclature called the pH is used to express the concentration of H^+ ions in a solution. The scale has 14 points and the midpoint, pH 7.0, is exactly neutral. A deviation from 7 towards 14 indicates increasing alkalinity, whereas a move from 7 towards 1 indicates increasing acidity. All pH values below 7 indicate acid solutions and above 7 they are alkaline. Each drop of one point on the scale means that the acidity has increased ten times. Blood at pH 7.4 is slightly alkaline. Urine varies from strongly acid, pH 5, to strongly alkaline, pH 8.

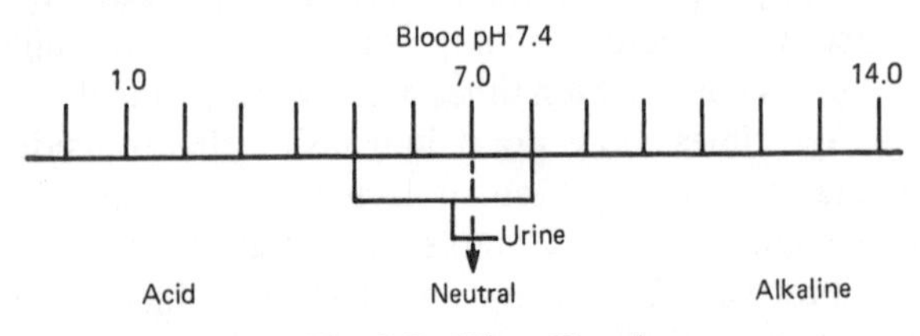

Fig. 2.5 The pH scale

The hydrogen ion concentration of the body is kept constant within very narrow limits by a series of processes. These are:

1 Chemical buffering;
2 Respiration;
3 Excretion by the kidney.

Chemical regulation of acid-base balance

The blood and tissue fluids contain chemical buffers which react with added acids and bases and minimise the resultant change in pH. Buffers are salts formed by the combination of strong bases with weak acids and the chief ones in blood are bicarbonate, haemoglobin and the plasma proteins, whereas in the urine phosphates perform a similar service. As stated above a strong acid contains a large amount of H^+ ion in solution and a weak acid contains small amounts. If a strong acid is added to a buffer it forms a new salt displacing the weak acid, which ionises so feebly that only a small rise in H^+ ion concentration results; in other words the strong acid is buffered.

Bicarbonate is normally present in blood and acts as a buffer to acids such as lactic, phosphoric, sulphuric and the fatty acids resulting in the liberation of carbonic acid, which is a weak acid and only produces small amounts of H^+ ion when in solution. Haemoglobin and the plasma proteins act as buffers in a similar way. The basic reaction is $H^+ + \text{Buffer} \rightleftharpoons \text{H-Buffer}$.

The acids which are constantly added to the tissue fluids by the cells are first buffered and then disposed of in various ways. Lactic acid resulting from muscular contraction is partly converted back into glycogen by the liver. Phosphates, sulphates and chlorides are excreted by the kidneys. Carbon dioxide, which is continually produced by all cells, is removed by the lungs, which thus exert an accurate and rapid control over the acid-base balance.

Respiratory control of acid-base balance

Carbon dioxide combines with water to form carbonic acid.

$$CO_2 + H_2O \rightarrow H_2CO_3$$

Although H_2CO_3 ionises feebly and only produces small amounts of H^+ ion it is present in large quantities in the body. Carbon dioxide can be removed very rapidly by increased breathing and therefore the carbon dioxide content of the blood can be quickly and effectively controlled. The respiratory centre in the medulla is extremely sensitive to the CO_2 content of the blood and probably also to the H^+ ion concentration, so an increase in either leads to overbreathing, which in turn reduces the concentrations of CO_2 and H^+ in the blood.

Renal regulation of the acid-base balance

The metabolic activities of the cells cause a constant release of acids into the tissue fluids, where they are first accommodated by the buffering mechanisms. The most important acids are lactic and phosphoric although sulphuric, uric and keto-acids are present in small amounts. Lactic acid, in the form of lactate, is dealt with by the liver (p. 48) but the other acid radicals are excreted by the kidneys in the form of an acid urine. Sometimes the extracellular fluid becomes too alkaline and the kidney then removes bases from the body by excreting an alkaline urine. The pH of the urine varies widely in these processes. When the tissue fluids are too acidic the pH of the urine may fall to 5.0 and when too alkaline it may rise to 8.0 but the pH of the blood and tissue fluids remains fairly constant within narrow limits.

3

The Living Cell

The cell is the fundamental unit of all animal life. All cells are constructed on the same basic pattern, although there are many individual variations. They consist of an outer cell membrane surrounding a jelly-like cytoplasm which contains many smaller units – organelles – and a central mass of protein and nucleic acids – the nucleus. Cells can perform these tasks:

1. Harness and transform energy. The cells of green plants change the energy of sunlight into chemical bond energy. Animal cells use foodstuffs as a source of chemical, mechanical or electrical energy.
2. Build larger molecules from small ones.
3. Make proteins which act as enzymes or form part of the structure of cells.
4. Reproduce themselves by division.
5. Develop into specialised cells with specific functions.

Structure of the cell

Cells are minute and only visible under the microscope except for a few, such as the ovum, which can be seen by the unaided eye. They are mostly 5–50 μm in diameter and a micrometre (μm) is only one-thousandth of a millimetre.

Each cell (Fig. 3.1) is bounded by a very thin membrane of protein and fat which maintains the chemical and electrical differences between the inside and outside of the cell. The membrane is

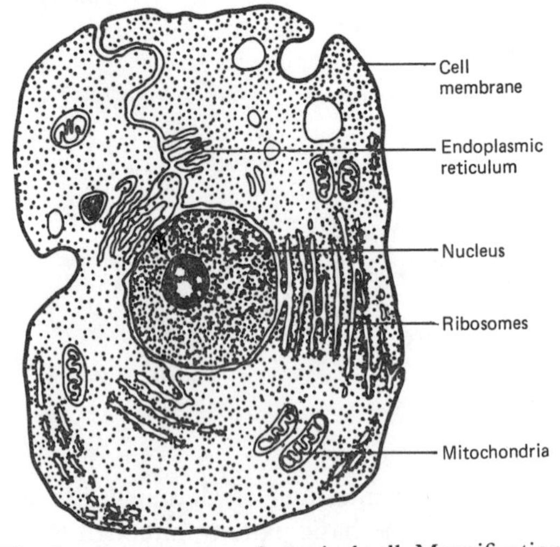

Fig. 3.1 The detailed structure of a typical cell. Magnification × 20 000

perforated by minute holes, which allow water and small molecules like oxygen, carbon dioxide and urea to pass freely through by diffusion. Larger molecules such as glucose, fatty acids and amino acids can only penetrate by active transport. They combine with a carrier molecule in the cell membrane, with the help of a specific enzyme. Glucose, for example, links up with phosphate to form glucose-phosphate, which is soluble in the cell membrane. At the inner surface of the membrane the glucose is split off and enters the cell. Larger particles are engulfed by a part of the membrane and taken into the cell as a tiny sphere or vacuole which travels to the appropriate region of the cell. This process is called *pinocytosis*.

The cell membrane forms the surface of contact with other cells. There is a very narrow gap between cells which is bridged by various special zones allowing contact between them. The membrane also carries the special tissue and blood antigens which are the markers unique to each individual. Infoldings and finger-like extensions (*villi*) of many cell membranes are associated with ion transport and the absorption or secretion of various substances.

Endoplasmic reticulum

This is a system of interconnecting membranous tubules or channels which permeates the cytoplasm of the cell and divides it into two compartments. The inner compartments store and transport the products of cellular activity. The outer spaces contain proteins such as enzymes, carbohydrates, amino acids and RNA molecules. There are two kinds of endoplasmic reticulum – rough and smooth. The rough areas contain the numerous ribosomes which synthesise proteins from amino acids while the smooth parts are concerned with other metabolic processes.

Ribosomes

These tiny bodies are made up of protein and a special form of *ribonucleic acid* (RNA) consisting of sugars, phosphates and bases linked together in specific patterns. RNA is similar to the *deoxyribonucleic acid* (DNA) of the chromosomes in the cell nucleus (p. 23). There are two related forms of RNA – messenger and transfer. The transfer RNA is found in the ribosomes and conveys the amino acids during protein synthesis. Messenger RNA passes out from the nucleus carrying the coded sequences of amino acids making up each specific protein.

The ribosomes are used by the RNA as 'assembly benches' where any required protein can be produced. The RNA picks up amino acids from the pool inside the cell in a pattern or sequence corresponding to the structure of the particular protein. The amino acids are carried to the ribosome, where they are linked together in the correct arrangement and proportion to produce the protein required. In this way the structural proteins of the cell and the enzymes, which are all proteins, are formed as needed.

Mitochondria

These sausage-shaped or oval bodies are from 0.5 to 2.0 μm wide and 3.0 to 4.0 μm long. There are from 2 or 3 to 700–800 per cell depending on the intensity of its metabolic activity. They have a smooth outer membrane separated by a gap from a highly convoluted or ridged inner membrane with a much greater surface area. The mitochondria contain ribosomes, nucleic acids and enzymes. They are the power plants where energy is extracted from the chemical bonds of the nutrients by the processes of oxidation and synthesis to form high energy compounds such as ATP (p. 11). The enzyme systems concerned with the citric acid cycle are also located in the mitochondria.

Lysosomes

These are smaller structures bounded by membrane and containing enzymes which break down proteins, carbohydrates and DNA. They are the digestive system of the cell and eliminate its worn out and degraded constituents. They can also kill and remove bacteria which have entered the body.

Golgi apparatus

This is a specialised zone of membrane near the cell nucleus which is most prominent in cells with a secretory function. It is very active metabolically and is the site of formation of the lysosomes, cytoplasmic and cellular membranes. Its enzymes prepare proteins formed elsewhere in the cell for secretion and are also concerned with carbohydrate synthesis.

Microtubules and microfilaments

These are minute cylinders and rods formed in many cells. They have a supporting, skeletal function giving stiffness and strength to the cell. They control the shape of cells and are an integral part of the structure of cilia. They are also concerned with cell division.

Cilia and flagella

These are tiny motile hair- and whip-like projections from the surface of cells, especially those lining tubes, as in the respiratory, reproductive and intestinal tracts. Cilia beat in unison producing waves which move adjacent material. Flagella cause movement by rhythmical bending like an eel in water.

Nucleus

Each cell contains a nucleus, which appears more densely stained than the rest of the cell when seen under the microscope. The nucleus is spherical with indentations and is surrounded by a nuclear envelope of membrane which merges with the endoplasmic reticulum of the cytoplasm. Pores act as conduits for the movement of molecules, such as messenger RNA, in and out of the nucleus.

The nucleus itself consists of a tangled mass of threads composed of nucleic acids and protein. Just before cell division (meiosis) they become visible as forty-six individual *chromosomes* arranged in pairs of different shapes and sizes. Twenty-two of the pairs are each structurally almost identical and are called autosomal chromosomes. The other pair determines the sex of the individual. In the female the two sex chromosomes are identical (XX) but in the male they are different, a single X chromosome being accompanied by a smaller Y chromosome (XY).

Every somatic (non-germ) cell in the body has the same full set of chromosomes containing the total hereditary instructions or blueprint for every cell in the body. In any particular cell only part of the total instructions are activated, thus determining the different cellular structures and functions in the various organs of the body.

Each chromosome carries a series of *genes* along its length. The genes for each inheritable characteristic are arranged in pairs (alleles) at exactly the same corresponding point on each pair of chromosomes. In any one chromosome pair the sequence of genes is stable and each cell in the body, apart from the germ cells, has exactly the same complement of genes. There are 20 000 to 90 000 genes in each chromosome pair and about 2 million in each cell. Many genes are structural but their activities are modified by various types of regulator genes. Structural genes determine the amino acid sequence of the enzyme proteins which determine the biological and structural features of each cell. The two alleles of a gene pair may be the same (homozygous) or may differ somewhat (heterozygous). If they differ they will have a dominant-recessive relationship to each other (p. 27). In asexual reproduction the genes are transmitted to the next generation in such a way that each new cell has exactly the same number and patterns of chromosomes and genes as the parent cell. In sexual reproduction half of each pair of chromosomes and genes is passed on to each daughter cell (pp. 25–6).

DNA consists of a very stable backbone made up of phosphate and sugar molecules joined together. At each join there is a side chain consisting of one of four bases – adenine, cytosine, guanine and thymine. The chemical characteristics of these bases force them to link together rather weakly in one way only – adenine with thymine and guanine with cytosine. Normally the DNA molecule consists of a double spiral with two backbones linked together by pairs of bases (Fig. 3.2). During cell division the spiral splits giving two strands of DNA which exactly match. One strand passes to each daughter cell so that the genetic information is passed on intact.

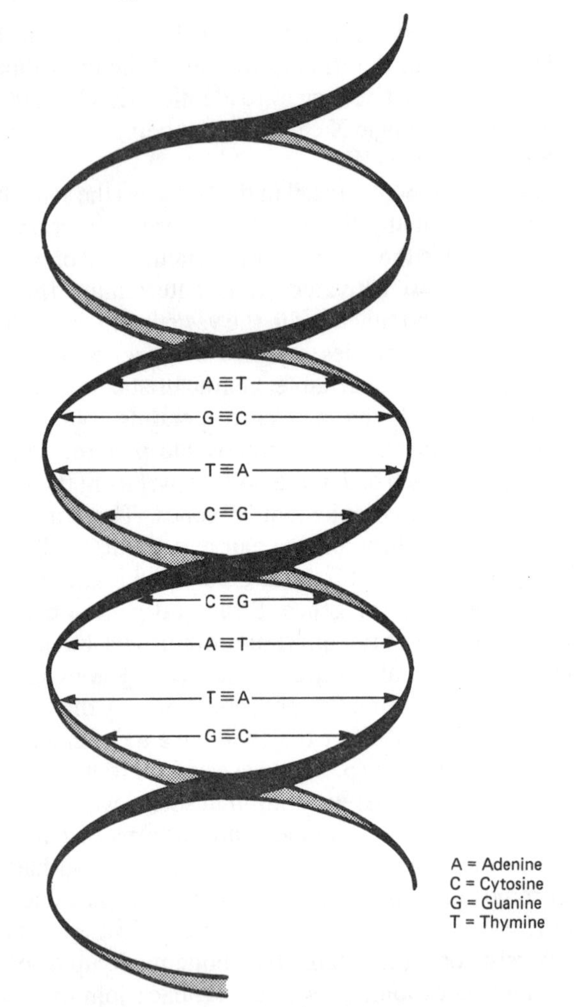

Fig. 3.2 The double helix of a DNA molecule

The arrangement of the bases along the backbone forms the genetic code. Each group of three bases corresponds to a particular amino acid and a sequence of bases can thus signify the amino acid composition of a particular protein. There are 64 possible triplets. Three of them are used to mark the end of each amino acid sequence. The rest are more than enough to encode the amino acid configurations of all the proteins needed. When a new protein is being assembled inside a cell the information carried by the genes is copied by segments of ribonucleic acid (RNA). RNA, which is similar to DNA but simpler, acts as a messenger and passes on the amino acid sequences to the ribosomes lining the endoplasmic reticulum. Here the structural proteins of the cell and its enzymes are constructed as needed. The genetic pattern of the fertilised ovum is transmitted to every one of the billions of cells in the adult body. Hereditable variations (mutations) in the structure of the chromosomes or genes may arise from errors during cell division or from the influence of chemicals or radiation. This can lead to profound changes in the structure and function of subsequent cell generations. Essential enzymes may be lacking causing disturbed function or disease. There may also be varying degrees of physical malformation. Many mutations are not compatible with survival.

Cell division

All the cells in the body divide by mitosis except for the germ cells, which undergo meiosis.

Mitosis

This begins with the doubling of the amount of DNA in the cell by synthesis. The nuclear envelope disintegrates and the chromosomes are released as separate dense rods. All chromosomes consist of two parallel and identical filaments – chromatids – joined together at some point by a centromere. The centromere splits and one chromatid of each chromosome moves to the opposite end of the cell. The cell then constricts down the middle forming two new cells. The nucleus of each cell develops a nuclear membrane and the DNA spreads out into a tangled network. In this way each daughter cell has exactly the same number of chromosomes as the parent cell and the same potentialities.

Meiosis

The process of division in the male and female germ cells is different. There is the same synthesis of DNA and the release of the pairs of chromosomes from the nucleus. The chromatids of each pair of chromosomes become partially entwined and there is some interchange of genetic material. The pairs of chromosomes then separate but the individual chromosomes remain intact. One chromosome from each pair moves at random to opposite poles of the cell. The cell constricts and divides into two daughter cells each containing 23 chromosomes instead of the normal 46. One of these cells contains little cytoplasm and is discarded. The other divides again without further reduction in the number of chromosomes but one cell is again discarded. There is thus a random assortment of chromosomes originally derived from either parent in this surviving ovum. When a sperm and the ovum fuse to form a fertilised egg the egg contains the full complement of 46 chromosomes, half the genes coming from each parent. This genetic re-allocation with recombination at fertilisation is the basis of inheritance and individual variation.

Determination of sex

The father's spermatozoa determine whether the baby is male or female. There are two special chromosomes related to sex in every body cell. In all female cells they are alike and can be designated XX, but in the male they differ, i.e. XY. When spermatozoa or ova are formed, one of each pair of the sex chromosomes is passed on to the germ cells. In females each ovum contains an X chromosome, but in males half the spermatozoa contain X chromosomes and half contain Y chromosomes. When fertilisation occurs there is an even chance that an X or Y loaded spermatozoon will be responsible and a similar chance that the baby will have XX sex chromosomes (female) or XY (male).

In addition to their primary function of producing the specialised cells necessary for reproduction, the sex organs (testis or ovary) secrete hormones which are essential for the normal functioning of the body. For convenience, these hormones are dealt with when the main theme of reproduction is described.

Inherited disorders

The genetic endowment of an individual is established at the time of conception but the way it unfolds and develops results from a constant interplay with environmental factors. It is probable that nearly all disease has some genetic component, if only a variation in susceptibility. At the other extreme there are diseases such as Down's syndrome (Mongolism) which are purely genetic. In between there are many diseases caused by a combination of genetic and external factors.

There are thousands of genes on a chromosome, each responsible for a single protein or enzyme. The genes are arranged in a permanent, fixed linear order which is identical on each matched pair of chromosomes. If the two genes at any point are not identical (heterozygous) the one which is expressed – becomes active – is called dominant and its partner is recessive. Recessive genes may be passed on to offspring but do not manifest themselves unless the individual possesses two of the recessive genes (homozygous). In plants the gene for tallness is dominant and that for shortness is its recessive. Heterozygous plants are always tall and plants homozygous for shortness are always short. Similarly in animals woolly hair is dominant and straight hair is recessive. If the gene is sex (X) linked the resulting trait will usually only appear in male or female, not in both. All other genes are autosomal. There are three types of inherited disorder:

1 Single gene mutations;
2 Multifactorial;
3 Chromosome abnormalities.

Single gene mutation

This means that a single amino acid is missing and there will be a biochemical change causing a metabolic disorder. The gene may be autosomal or X-linked and this determines the pattern of inheritance. In phenylketonuria an enzyme which normally converts the amino acid phenylalanine to tyrosine is deficient and permanent mental retardation results. If detected early enough, this can be countered by a special diet which excludes phenylalanine. Colour blindness is due to an abnormality in one of the retinal visual pigments (p. 170) and the gene is X-linked. The sufferers are usually male but if the genes are the same (homozygous) in a female, she will be colour blind. In haemophilia there is an abnormality of blood clotting due to an X-linked gene. As a result the patients are nearly always male but the disease is transmitted by any female carrying the appropriate gene. Very rarely a female haemophiliac occurs because she is homozygous for the gene.

Multifactorial

This arises from a combination of several gene defects, each causing a minor anomaly, interacting with several different environmental factors. Examples of this are spina bifida, congenital heart disease and cleft palate.

Chromosomal abnormalities

Using special techniques it is possible to study the individual chromosomes microscopically and in certain conditions structural abnormalities can be seen. In Down's syndrome an extra chromosome is present and the individuals have a characteristic appearance with mental retardation. They are often afflicted with congenital heart disease and are liable to develop acute leukaemia. In other cases there may be one, two or three extra X chromosomes producing a series of rare congenital abnormalities.

Prevention of genetic disorders

Genetic counselling based on family history and examination of existing children is the basic approach leading to the prevention of the conception of abnormal foetuses. During pregnancy a little fluid from the uterus can be drawn off and chromosomal or biochemical anomalies can be detected, allowing for abortion if indicated. Newborn infants can be screened for conditions such as phenylketonuria and appropriate management instituted.

Blood groups

The red blood cells (erythrocytes) can be differentiated by the presence or absence of glycoprotein markers on their surface. These markers act as antigens (p. 85) and react with specific antibodies. The clinically important groups are the ABO and Rhesus (Rh) systems although there are many others found on special testing. The ABO system depends on the presence or absence of two allelic genes A and B and is described on page 85. It is vitally important in relation to blood transfusion. The Rh system is crucial in some pregnancies and is dealt with on page 86.

The HLA system

A complex of genes on the sixth chromosome determines the presence or absence of a group of antigens on the tissue cells. They are tested for on human leucocytes (pp. 82–3), hence HLA. They are the basic reason for the rejection of foreign tissue transplants such as skin, heart, liver and kidney. This can be partly overcome by careful matching of the HLA systems in donor and receiver. There is also statistical evidence that certain diseases of the skin, kidneys and joints are commoner in individuals with particular HLA types.

4

The Tissues of the Body

In the very early stages of development the embryo consists of three layers, an outer protective *ectoderm*, an inner nutritional *endoderm* and an intermediate layer of *mesoderm*. The ectoderm forms the skin, hair, nails, nervous system and parts of the eyes. The endoderm gives rise to the linings of the air passages, renal tract and gut plus the glands opening into it such as the liver and pancreas. From the mesoderm come the muscles, the connective tissues, the linings of the chest and abdomen, the blood vessels and the lymphatic system.

Types of tissue

There are four basic tissues, epithelial, connective, muscular and nervous. *Epithelium* can be derived from any of the three embryonic layers. It consists of sheets of cells covering or lining the surfaces of the body and forms a protective barrier with one or more layers of cells resting on a basement membrane (Fig. 4.1). It also forms the functioning parts of the glands. Epithelium regenerates very quickly after injury as is seen in the damaged liver and in the skin, which is continually being worn away and replaced. The epithelium lining the gut, air passages and renal tract contains many glands which secrete mucus. Such mucous membranes protect against drying out, provide lubrication and trap foreign matter such as dust. The mucus is a sticky, viscous suspension of glycoproteins.

Connective tissue (Fig. 4.2) is the main supporting material and consists of cells widely spaced in an intercellular *matrix* produced by

Cells are often arranged in sheets that cover or line organs.
Some examples are:

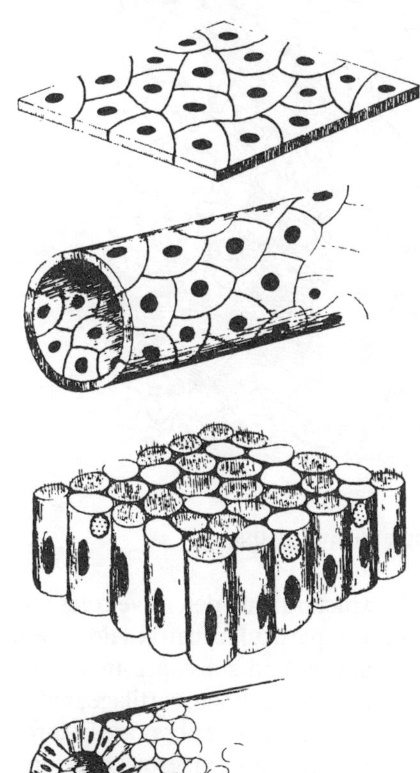

A thin sheet of flattened cells forming a smooth surface for an organ - the *pleura* over the lungs.

A rolled-up thin sheet forming a smooth lining for a tube - the inner lining of an artery.

A thick sheet formed of a single layer of tall columnar cells - some producing globules of sticky mucus and others bearing small moving cilia which brush material over the surface - the mucous membrane lining the nose.

A rolled-up thick sheet of cells which lines a tube and pours mucus into it - the lining of the last part of the food tube (the *rectum*)

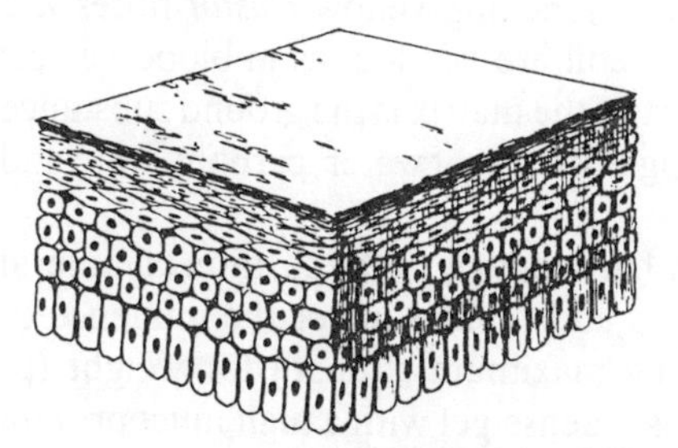

A many-layered sheet with flattened dead cells on the surface that can be worn away harmlessly - the *skin*

Fig. 4.1 Cell patterns

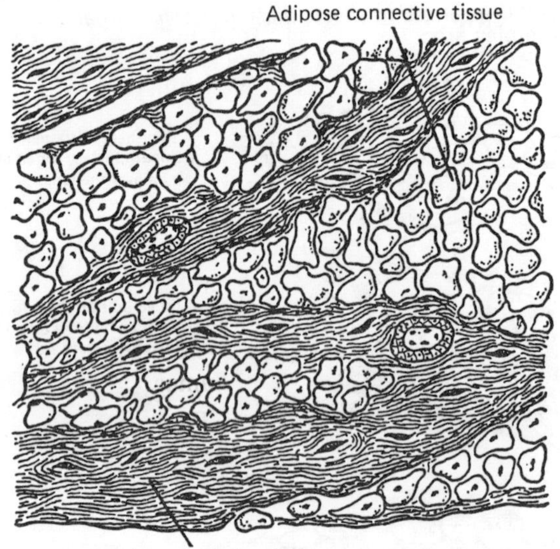

Fig. 4.2 Simple connective tissues

the cells themselves. It has both structural and defensive roles and its properties depend largely on the particular matrix laid down. The matrix is formed of fibres and ground substance in various proportions. Elastic, fibrous and fatty tissue, bone, cartilage, blood and lymph are all forms of connective tissue. The fibres are of three types. The white *collagen* fibres are the most numerous and form inextensible bundles in tendons, ligaments and fibrous structures.

Reticulin fibres are a thin, branching, anastomosing form of collagen providing a supporting framework between the collagen bundles and in many glands. Branching yellow *elastin* fibres are found wherever stretch and recoil are needed, as in blood vessels and skin. The non-fibrous part of the matrix is the ground substance which is a viscous gel with a high content of water, carbohydrate and protein.

In bone, collagen fibres are hardened by the deposition of mineral salts and organised into layers and tubules under the influence of external forces thus providing maximum strength for weight (p. 36). The matrix of cartilage is a dense gel with a high mucoprotein and polysaccharide content in which are embedded cartilage cells and a stiff feltwork of collagen fibres. There are few blood vessels supplying connective tissue and its metabolic needs are largely met by diffusion with drainage into the numerous lymphatic vessels which permeate most connective tissues. Nerve fibres and endings are also relatively scanty. The cells of blood and lymph have developed special properties for the transport of gases and in defence. They float in a completely liquid matrix.

The most numerous cells in connective tissue are *fibroblasts*. They produce the collagen fibres and are specially active after injury and tissue damage. *Macrophages*, both fixed and motile, are also numerous. They are able to engulf and digest (by phagocytosis) foreign particles such as bacteria or damaged tissue. They are part of the reticulo-endothelial system (p. 86). *Mast cells* are less common and their function is not clear. They contain histamine, serotonin and heparin and are probably concerned with bodily responses to infection, injury and with hypersensitivity reactions. Lymphocytes and plasma cells move into the connective tissues as a part of the immunological defence mechanisms (pp. 89–90).

Most effective human activity results from muscular contraction. *Muscles* are specialised tissues which convert chemical energy into power and motion. There are two main kinds of muscles in the body, each with different characteristics but with basically similar properties. They are:

1 (*a*) Skeletal – striated or striped muscle.
  (*b*) Cardiac – a special form of striped muscle.
2 Visceral – unstriated, unstriped or smooth muscle.

Striped muscles are under the control of the central nervous system and most skeletal muscle can be consciously and deliberately contracted. There are several exceptions to this, for example, the intercostal muscles between the ribs and the striped muscle in the floor of the pelvis. These muscles can only be contracted as part of more complex activities, the intercostals in breathing and the pelvic floor muscles in the act of defaecation or in association with voluntary contraction of the external anal sphincter and buttock muscles. Smooth muscle is under the control of the autonomic nervous system (p. 159) and cannot be voluntarily controlled although it has considerable intrinsic contractile properties.

The *nervous system* is the most highly developed and organised in the body. It controls and supervises the activities and functions of the whole organism, which would not survive without it. It enables swift communication and interaction between the tissues and organs in all parts of the body. It organises and initiates the muscular contractions which are the final expression of all effective human behaviour. It consists of an input or *receptor* system, a central recording and coordinating complex and an *effector* network of motor nerve fibres.

Tissue fluid

An adult male body of 70 kg (11 stones) contains 40 litres (10 gallons) of water. About 25 per cent of this (11 litres) is outside the cells. This *tissue fluid* permeates the whole body and is separated from the intracellular fluid by the cell membranes and from the blood plasma by the capillary walls. The special properties of the cell membranes maintain important differences in chemical composition between the intracellular fluid and the tissue fluid (p. 14). Tissue fluid is the medium of transport of all foods, gases and waste products between the cells and the blood plasma.

The chemical composition of tissue fluid and blood plasma (p. 80) is essentially the same, except that there is relatively less protein in the tissue fluid. An increase in the amount of waste products in the tissue fluid, or a decrease in the amount of food or oxygen, is rapidly corrected by an increase in the amount of blood flowing through the affected part of the body. There is a continuous exchange of water and chemical substances between the blood plasma and the tissue fluid. Normally, the tissue fluid is renewed every few minutes, but in active organs the capillaries are dilated and the turnover of tissue fluid is much faster.

Arterial blood enters the capillary vessels under a pressure equal to that exerted by a column of mercury 35 mm high. This hydrostatic pressure tends to force water molecules and all chemical particles smaller than the plasma proteins out through the porous walls of the capillaries into the spaces between the cells (Fig. 4.3). The plasma proteins are too large to escape and they exert an *osmotic pressure* inwards of 25 mm of mercury. The net outward pressure at the arterial end of the capillary is 10 mm of mercury. At the venous end of the capillary the hydrostatic pressure has fallen to 15 mm of

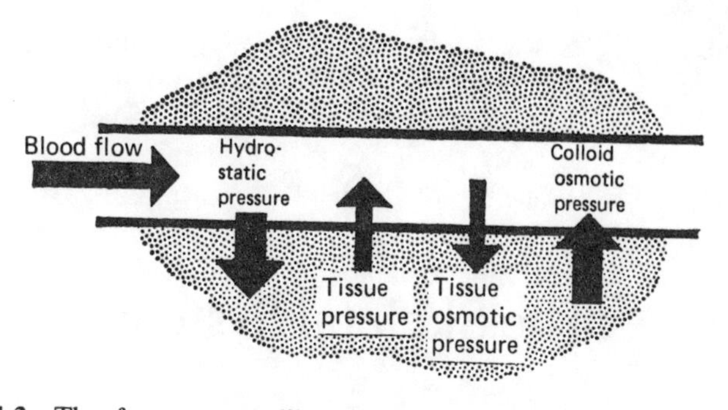

Fig. 4.3 The forces controlling the movement of water between the capillaries and the tissue spaces

mercury, but the osmotic pressure of the plasma proteins now exerts a net inward force of 5–10 mm of mercury which attracts tissue fluid back into the capillaries.

5

The Skeletal System

The skeleton consists of a supporting and protective framework of bone and cartilage linked by joints and ligaments. The skeleton is integral with its muscles and they will be described together. They are exquisitely adapted to their functions in the motor activities of the individual. The shape, size and structure of each component part is the resultant of genetic, metabolic and mechanical factors. Sexual and racial differences are genetically determined as are, to a considerable extent, shape, size and strength. Metabolic influences are continually operative. The supply of calcium, phosphorus, vitamins A, C and D and hormones from the thyroid, parathyroid, adrenal and pituitary glands, plus the gonads, are all critical. The mechanical effect of muscular activity is also crucial as can be seen in the deformity of a limb paralysed by poliomyelitis.

Types of bone

One-third of bone is water yet it has the tensile strength of cast iron. It develops from cartilage and consists of a matrix of mucopolysaccharide containing a dense feltwork of collagen fibres made rigid by the deposition of tiny crystals of a compound of calcium and phosphorus. The bone is moulded and organised by internal and external forces into minute tubules and layers. Throughout life its infrastructure is continually modified by *osteoclast cells*, which erode and remove the lamellae, and by *osteoblasts*, which lay down fresh matrix.

There are three main types of bone – long, short and flat. The large long bones are found in the limbs. They have a round or oval shaft and a cartilage covered expansion at each end forming half of a neighbouring joint. They have an outer layer or cortex which is weight-bearing and a vascular marrow in the centre. In many bones this is the seat of formation of the red blood corpuscles. Small long bones are found in the hands, feet, fingers and toes.

Short bones are squat, irregular on circular blocks. They have a thin solid cortex with a spongy bony network inside. They are found in the vertebrae of the spine, the wrists and the ankles.

The flat bones consist of two plates of compact bones with a thin spongy interior. They form the vault of the skull, the ribs and the shoulder blades (scapulae).

Types of joint

The contact surfaces between bones produce various types of joint with different degrees of mobility and stability (Fig. 5.1). The main types are:

1 *Fibrous*. The bones are joined by layers or sheets of fibrous tissue allowing little movement. These joints are found between the bones of the skull or between paired bones like the radius and ulna.
2 *Cartilaginous*. Here there is a thick plate of fibrocartilage joining the two bones as in the discs between the vertebrae or the sacro-iliac joints between the two halves of the pelvis and the sacrum.
3 *Synovial Joints*. These allow considerable degrees of movement. The ends of the bones are covered by smooth cartilage and the joint is enclosed by a fibrous capsule. The joint is stabilised and supported by strong fibrous ligaments with the help of muscular insertions. The capsule is lined by a very vascular synovial membrane which secretes the clear, yellow, slippery synovial fluid. This is a filtrate of blood plasma very similar to tissue fluid but containing a polysaccharide – *mucin* – which lubricates the joint. The articular cartilage contains neither blood vessels nor nerves and is therefore insensitive, but the synovial membrane is richly supplied with both. Nerve

the fibrous joints

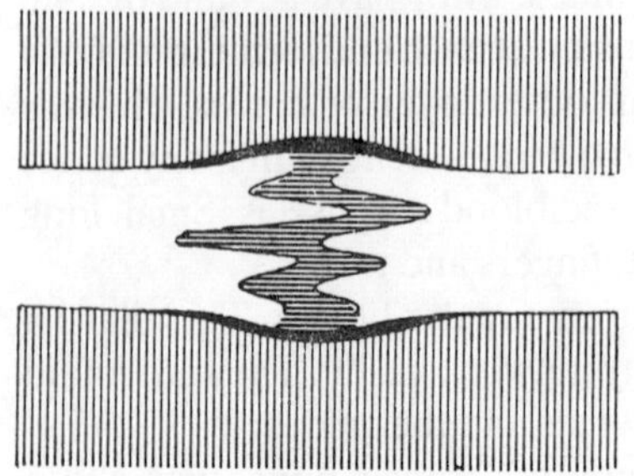

The bones of the skull are united by fibrous joints that allow no movement at all.

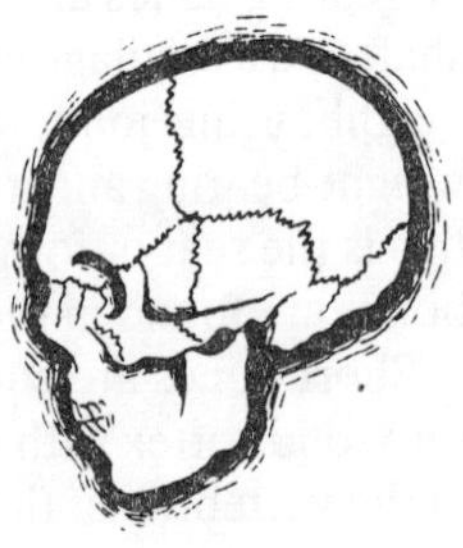

the cartilaginous joints

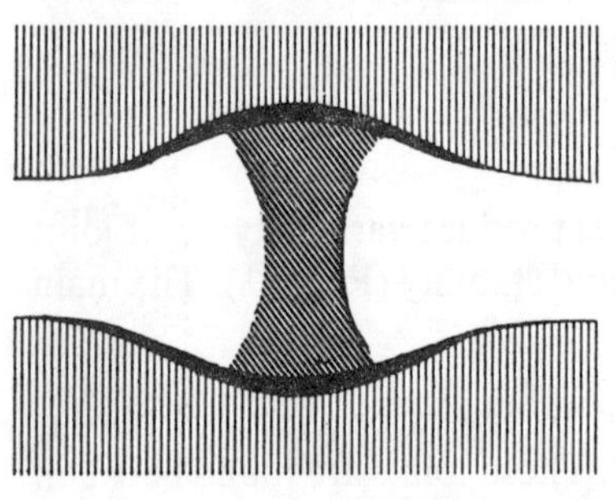

The vertebrae are linked mainly by cartilaginous joints between the vertebral bodies.

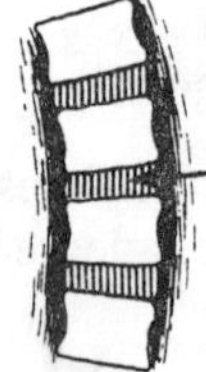

These 'discs' act as shock absorbers as well as allowing one vertebra to tilt or rotate on the next.

the synovial joints

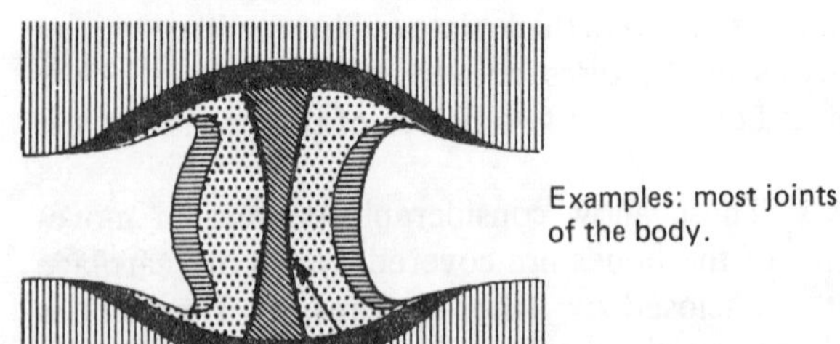

Examples: most joints of the body.

Usually there are thick bands of white fibres - the ligaments - assisting the capsule.

In a few joints there is a disc of fibro-cartilage here with a synovial cavity on each side.

Fig. 5.1 Types of joint

impulses transmitting information about the position, movement and state of the joints arise from nerve endings in and around the joint but not in its articular surfaces.

There are several types of synovial joint with different capacities and ranges of movement. Hinge joints are found at the elbow, ankle and fingers. A saddle joint is seen between the thumb and wrist. The widest range of movement is found in the ball-and-socket joints at the shoulder and hip. The knee is the largest joint. It is a compound joint with three separate articular surfaces and two pads of fibrocartilage inserted between the weight-bearing surfaces. It permits both flexion–extension and rotation. Its complexity makes it very liable to injury at work or in sports.

The bony skeleton

The skeleton gives support and protection to the delicate and vulnerable soft tissues, and also provides firm anchorage for the muscles (Fig. 5.2). With its joints and levers it is efficiently developed to provide the basis for nearly all muscular activity ranging from the lifting of heavy weights to the skills of the craftsman and the artistry of the pianist.

The skeleton consists of an axial portion, the head and trunk, and the appendages of the limbs. The skull is a box of interlocking plates of bone protecting the eyeballs and brain which is balanced upon the spine. This is a column of bony blocks – *vertebrae* – which form a strut for the trunk and a bony, protective tube for the spinal cord. The spine is flexible because each vertebral body is joined to its neighbour by a thick disc of fibrocartilage. The thoracic cage consists of twelve pairs of semicircular bony ribs, joined to the breast bone (*sternum*) in front by strips of cartilage (Fig. 5.3). It protects the heart and lungs and also shelters some of the upper abdominal organs. There are no bones, except the spinal column, inside the muscular wall of the abdomen but the lower abdominal organs are shielded by the flat bones of the *pelvic girdle*. This consists of five vertebrae fused to form the *sacrum* and the sturdy semicircles of pelvic bones which join in front. The pelvis transmits the body weight to the legs and supports the whole body when sitting.

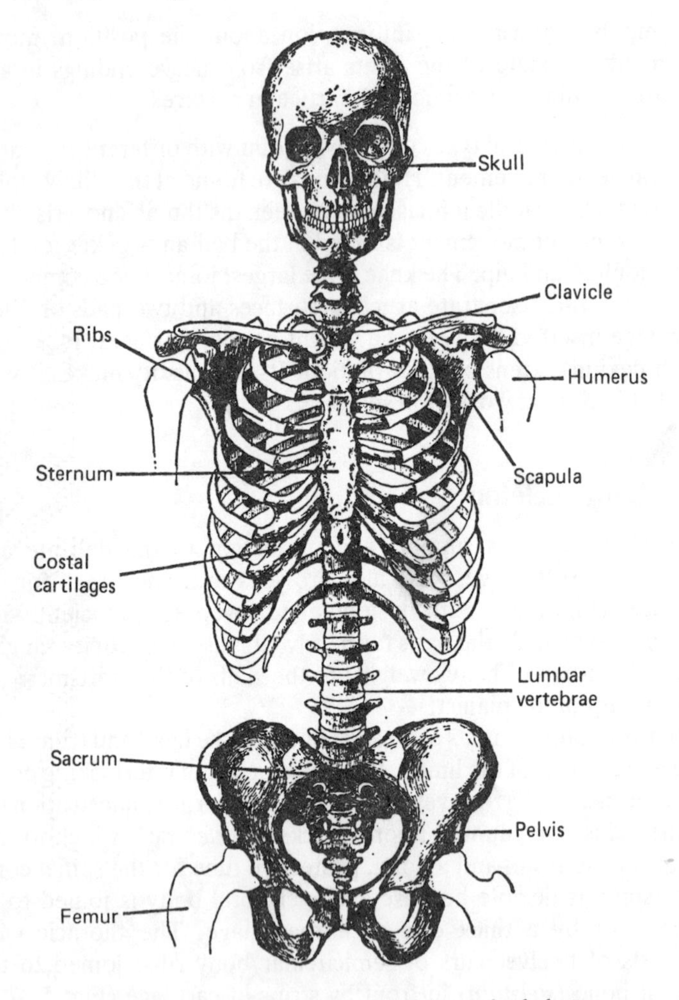

Fig. 5.2 The head and trunk of the male skeleton

The two pairs of limbs are built on a standard pattern with struts along their whole length. The shoulder blade (*scapula*) is joined to the breast bone by the collar bone (*clavicle*) and slides over the back and side of the chest. It is attached to the *humerus* by a ball-and-socket joint which allows free movement. The forearm (Fig. 5.5) contains two bones, the *radius* and *ulna*, that join up with a

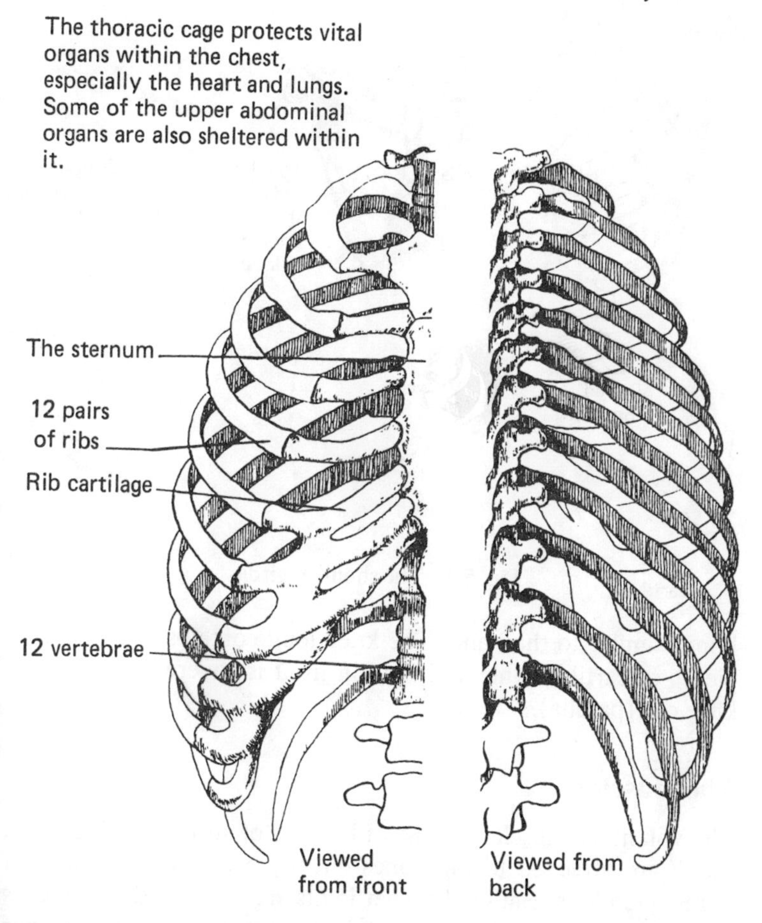

Fig. 5.3 The thorax

complicated cluster of wrist bones, which in turn support bones of the hand and fingers. The elbow joint is a hinge formed by an upward projection of the ulna – the *olecranon*. The leg has a similar pattern of bones. The thigh bone (*femur*) is inserted into a ball-and-socket joint at the side of the pelvis. Its lower end forms a weight-bearing hinge joint with another large bone of the leg – the *tibia*. On the outer side of the tibia a slender bone, the *fibula*, runs down to the ankle joint. The ankle and foot have a bone pattern of small

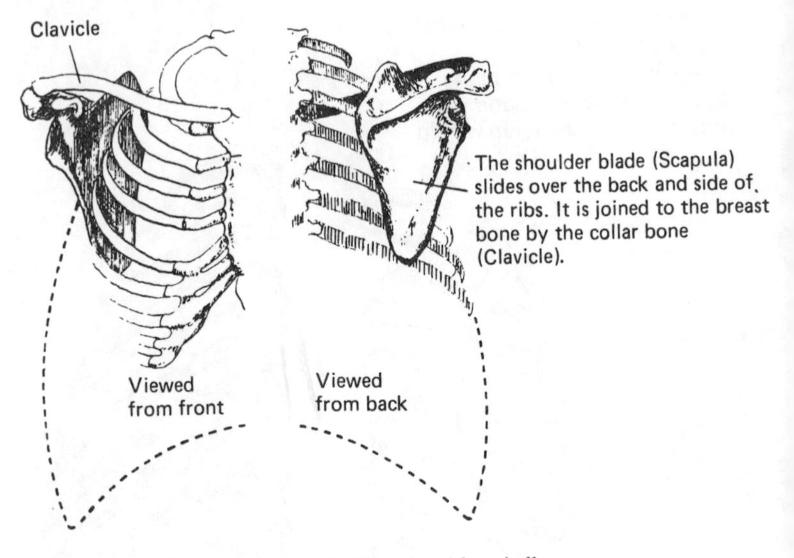

Fig. 5.4 The shoulder girdle

bones similar to the hand. The knee joint contains two C-shaped pieces of cartilage (menisci) and in front is covered by the bony knee cap (patella).

The muscles

The voluntary muscles consist of bundles of striped muscle fibres which are attached to the bones directly or by stout bundles of fibres – tendons. Ball-and-socket joints are controlled by bunches or clusters of muscles which hold the joint surfaces firmly together yet allow a wide range of movements. Hinge joints are moved by simple muscle groupings arranged to straighten or bend the part. The triceps muscle at the back of the upper arm straightens the elbow joint and is the antagonist of the biceps muscle in front, which bends it (Fig. 5.6). The forearm contains an intricate pattern of interlaced muscles which control the movements of the wrist and fingers through very long tendons. Much of the precision of hand movements derives from the numerous small muscles running between the small bones of the wrist, palm and fingers.

Similarly, the big quadriceps muscle in the front of the thigh

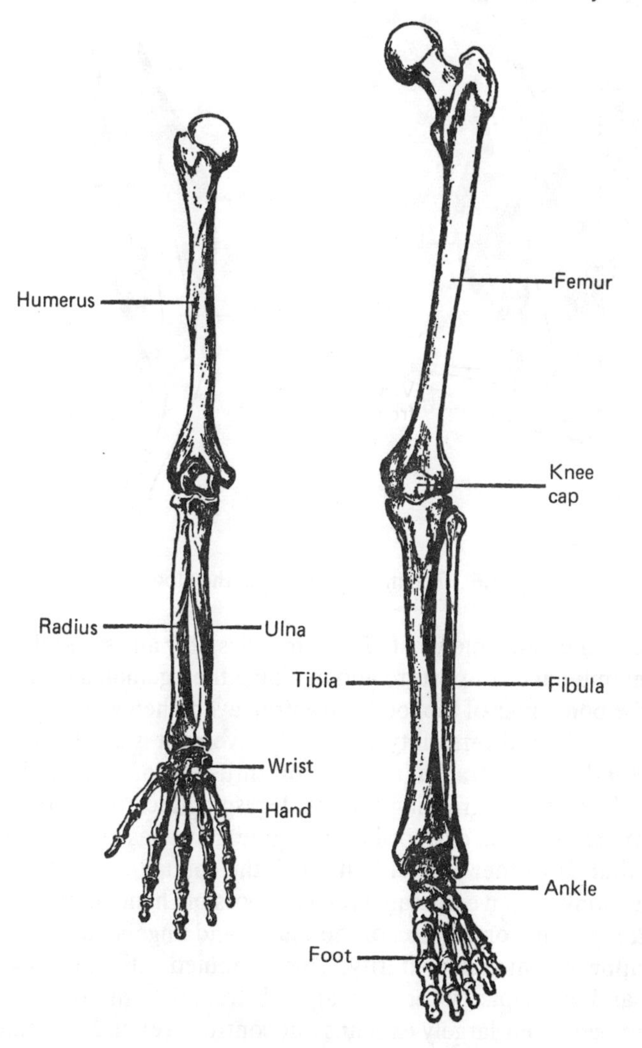

Fig. 5.5 The bones of the arm and leg

straightens the knee joint and is the antagonist of the group of hamstring muscles which bend it. The calf contains the *gastrocnemius* and *soleus* muscles which act upon the foot and ankle joint in opposition to the *tibial* and *peroneal* muscles which lie in front. The foot and ankle contain numerous small muscles comparable to

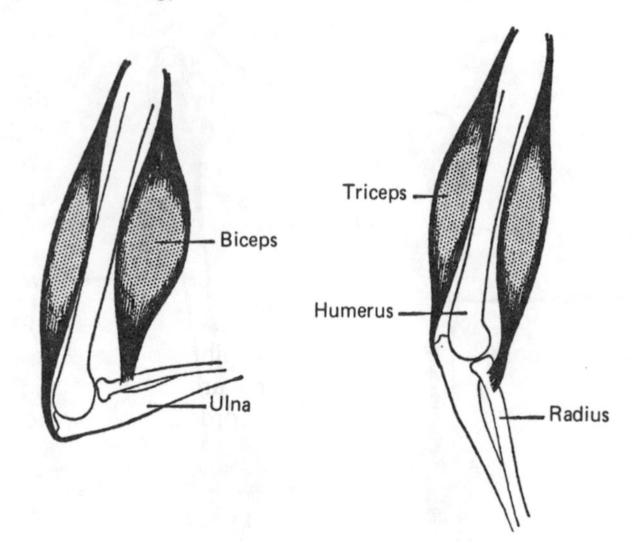

Fig. 5.6 The muscles acting at the elbow joint

those in the hand and wrist. These muscles play an essential part in the maintenance of posture and in aiding the ligaments in converting the bony links of the foot into a firm lever when walking.

The range and versatility of skilled movement is much greater in the hand. It can transmit power by thrusting or striking but its special value is in grasping actions. Grasping includes both power and precision to an extraordinary degree. Man has greater manual skill than any other animal. Although the chimpanzee, gorilla and some monkeys have the same range of possible hand movements as man, especially opposition of the thumb and fingers, none of them can approach man in dexterity. This combined with innate aggression and our capacity for language, abstract thought and directed group behaviour largely explains our control over and exploitation of our environment.

The muscles and tendons of the abdominal wall form flattened sheets, which are arranged in layers like plywood to give maximum support. The outer and middle layers run obliquely and the inner layer is transverse. In front there are two powerful bundles of muscles running vertically from the pelvic girdle to the rib cage on each side of the midline – the *rectus abdominis muscles*. The thoracic cage is separated from the abdomen by a thin curved sheet of muscle and tendon – the *diaphragm*. Its muscle fibres radiate from the flat central tendon to be attached to the inner surface of the chest wall. When they contract in breathing they pull the tendon downwards, enlarging the chest and depressing the abdominal organs. The diaphragm separates the chest from the abdomen but is pierced by the main artery and vein and by the oesophagus.

Muscle attachments

Tendons are virtually inextensible, integral parts of the muscles. They are composed of bundles of collagen fibres running in parallel. Their flexibility enables them to bend around bone and joint prominences or to run through tunnels in order to change their final direction of pull. They are smooth but pale because of their scanty blood supply. At many of these deflection sites *bursae* are formed. These are sacs of synovial membrane containing a little fluid to ensure easy movement. In addition *synovial sheaths* form around tendons where they pass beneath fibrous bands as at the wrist or ankle.

In all muscles the contractile fibres are interspersed with collagen fibres which support them and distribute the load. These are arranged in flat, dense sheets or *aponeuroses* and in a variety of more delicate layers and bundles forming superficial and deep fascia. Each of the separate muscle fibres in a muscle is wrapped in a thin sheath of delicate connective tissue, and each group of muscle fibres forms a *fasciculus*, which is a bundle of fibres with a tougher connective tissue wrapping. Several fasciculi are caught up in connective tissue to form secondary and tertiary bundles which, in aggregate, make up the whole muscle. The interconnecting meshwork of connective tissue throughout the muscle blends with the tough fascia covering it and forms a framework which is attached to tendon or bone and transmits the power from the muscle fibres (Fig. 5.7).

In short muscles the fibres may run the whole length but in longer ones the fibre bundles form chains running from one end to the other. In other muscles the fibres and fibre bundles run obliquely and are inserted into central tendons and fascial sheets. Such muscles are more powerful because they contain a larger number of fibres per unit length.

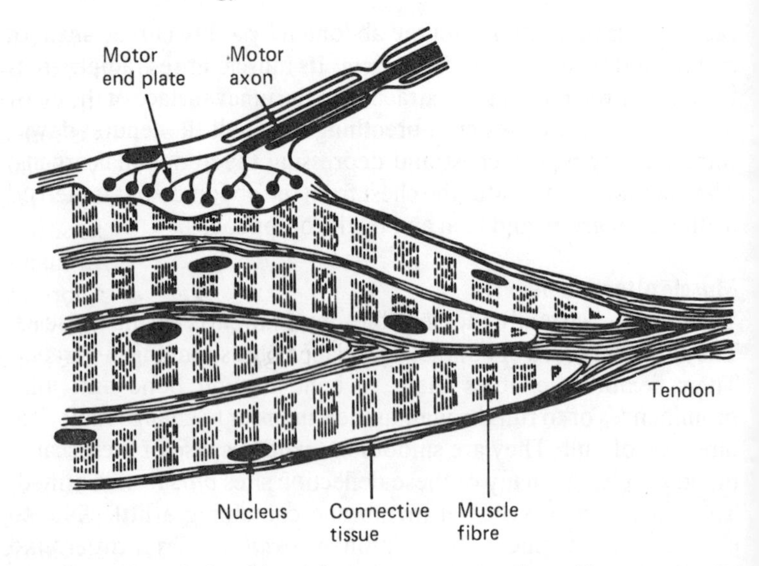

Fig. 5.7 The relationship of the motor nerve endings, fibres, connective tissue and tendon insertion of a muscle. Structures highly magnified, schematic and not to scale

Muscle spindles

The efficient performance of a highly skilled activity such as playing the piano would not be possible unless the central nervous system were continuously informed of the state of contraction of the muscles under its control. All skeletal muscles contain specialised sense organs – muscle spindles – which are sensitive to changes in tension in the muscle. They lie parallel to or alongside the muscle fibres, so that if the muscle is stretched they are stimulated and an increased number of nerve impulses pass up the nerve fibres which connect them to the central nervous system.

If the muscle itself contracts then the pull upon the muscle spindle diminishes and the number of impulses travelling centrally is reduced. The sensory nerve fibres from the muscle spindles terminate, partly, on the anterior horn cells which supply the motor units of the same muscle and thus can modify their excitability and rates of firing. In this way the contraction of the muscle is constantly monitored and further contraction is exactly graded to the requirements of the situation and to the tension which the muscle has already exerted. The muscle spindles are normally very sensitive to changes in tension, but their sensitivity or setting can also be modified by nerve impulses which travel down to them from the central nervous system in special small nerve fibres. This is called the *gamma efferent system* and the nerve fibres end on tiny specialised muscle fibres inside the muscle spindles themselves. The contraction of these tiny muscle fibres alters the setting or sensitivity of the muscle spindle so that it fires to more or less tension than is normally the case. The muscle spindle and its sensory nerve fibre are the input or *afferent* side of the reflex response which occurs when the muscle is suddenly stretched by tapping its tendon, e.g. the knee jerk.

Fast and slow muscle fibres

One limiting factor in muscular performance is the speed of contraction of the muscle. When a muscle exerts its maximal force the speed of contraction is less. A heavy weight can be lifted only slowly but a light weight can be thrown at great speed. The trained athlete employs the most economical choice of force and speed of contraction to achieve the required object. Natural athletes achieve this economy of motion by intuitive processes, but nearly all athletes can improve their performance by skilled coaching which enables them to increase the efficiency of their muscular activity.

Another factor may be the relative proportions of fast and slow muscle fibres. These have been clearly demonstrated in most experimental animals as well as in man. Most muscles show a mixture of slow (Type 1) and fast (Type 2) fibres. They differ in appearance and in their metabolic processes. Type 1 fibres contract with a twitch lasting about 75 msec. They are richly supplied with mitochondria containing oxidative enzymes that enable them to respire aerobically. They are adapted to the slower, more sustained activity of the postural muscles. Type 2 fibre contractions last about 25 msec. They have few oxidative enzymes and utilise anaerobic recovery processes. They are better suited to the rapid phasic contraction of skilled, voluntary activity.

Structure of skeletal muscle

There are about 400 skeletal muscles made up of about 250 million cells called muscle fibres. They comprise about 30 per cent of the total body weight. Muscle fibres are elongated structures, 1–50 mm long but only 10–60 μm in diameter, which are attached at both ends to either connective tissue or bone. The cell membrane around each muscle fibre is called the *sarcolemma* and numerous flattened nuclei lie underneath it. There are many ribosomes and large, elongated mitochondria. Each fibre is full of minute protein threads (*myofibrils*), 1 μm in diameter. The myofibrils run along the whole length of the muscle fibres and are made up of many myofilaments which have the property of shortening by infolding upon themselves. They consist of two types of protein threads – thin *actin* and thick *myosin*. They are arranged in alternating bundles which interlock with each other along the myofibril. The striations of skeletal muscle seen under the microscope are produced by the dark bands of thick myosin alternating with the light thin bands of actin. When a muscle fibre contracts in response to a nerve impulse a wave of depolarisation sweeps down the sarcolemmal membrane liberating calcium ions. These activate the myosin and actin molecules so that they interlock with each other and move closer together thus shortening the fibre. The power of muscular contraction results from the forces exerted by the filamentous protein molecules when they shorten and overlap. The immediate energy for muscular contraction comes from the breakdown of energy-rich adenosine triphosphate (ATP) to form ADP (p. 11). The ATP is quickly reformed from other chemical bonds in the muscle fibres which are later replenished in turn by the oxidation of fatty acids and glucose with the production of carbon dioxide and water. If the oxygen supply is insufficient energy is obtained from the anaerobic breakdown of glycogen and an excess of lactic acid is formed. The lactic acid diffuses out of the muscle into the blood stream and is carried back to the liver, where it is re-formed into glycogen. The liver glycogen is the source of much of the sugar in the blood from which the muscle glycogen is again formed. The cycle, muscle glycogen → blood lactate → liver glycogen → blood glucose → muscle glycogen, is essential for the supply of energy.

Oxygen debt

The recovery processes after muscular contraction require an adequate supply of oxygen which is conveyed to the muscle from the lungs by the blood stream. In moderate muscular exertion requiring up to 2 litres of oxygen per minute, a steady state can be achieved in which recovery keeps pace with the chemical changes in the muscle. When more severe muscular exertion is undertaken the chemical processes can continue to a certain extent in the absence of oxygen, but an 'oxygen debt' is incurred which has to be repaid later.

When muscle contracts without adequate supplies of oxygen the breakdown of glycogen forms lactic acid in excess. The re-synthesis of lactic acid into glycogen requires oxygen and when muscular exertion outstrips the oxygen supply the amount of lactic acid in the body increases steadily. In this way lactic acid equivalent to 10–15 litres of oxygen can accumulate, but it has to be dealt with when the exertion is over. Respiration and oxygen intake then occur at rates far above the normal resting level for the subject. This is the reason for the prolonged overbreathing which follows any severe muscular exertion.

Muscle blood supply

Arterial blood is conveyed to the muscles through the branching arteriolar system, which terminates in very large numbers of capillaries infiltrating throughout the muscle between the individual muscle fibres. The exchange of gases, fluids and nutrients occurs at the capillary level. The blood supply to the resting muscle is small but it rapidly increases when the muscle begins to contract. During exertion the cardiac output and rate of circulation increase markedly (p. 132) and more blood passes through working muscles, in which the arterioles and capillaries are dilated. The dilatation of the intramuscular vessels is partly due to the liberation of adrenaline by the suprarenal glands into the blood stream, which occurs in association with exertion, and partly to the local vasodilator effects of the metabolites which accumulate in the muscle during exertion. These mechanisms can increase the muscular blood flow tenfold or more during exercise, but there is an upper limit to the blood flow through the muscle which is related to the force or tension exerted by the muscle during contraction.

During a full sustained contraction, e.g. standing on tip-toe, the intramuscular pressure rises so high that the blood flow is cut off, but usually muscles alternate between contraction and relaxation and the latter allows the blood to surge through them. After exercise there is a prolonged increase in the blood flow above the resting level for the muscle. This is due to the vasodilatation resulting from the local action of the metabolites which accumulate in the muscle. This reactive hyperaemia continues for many minutes after prolonged muscular activity until the metabolites have been removed.

The contraction of muscle

A muscle fibre is normally in the relaxed or resting state but when an adequate stimulus, i.e. any form of mechanical, chemical or electrical energy, is applied, it will contract momentarily. If the muscle fibre is fixed at both ends the contraction will produce an increase in tension in the muscle without any shortening in the fibres, as in standing or pushing against a wall. Such a contraction is called static or *isometric*. If only one end of the fibre is tethered it will shorten, as in lifting a weight. Such a contraction is called dynamic or *isotonic*.

A muscle fibre will not contract unless a sufficient amount of stimulus energy is applied to it, but eventually it either contracts fully or not at all. Increasing the energy of the stimulus above this level does not increase the force of the contraction and this is termed the all-or-none law of muscular contraction. It is comparable with the all-or-none law of nervous transmission (p. 136) but there are some differences.

The excitability of a muscle fibre varies from time to time, like that of a neurone, in relation to its previous activity, but it is much more susceptible to fatigue than is a neurone. Its contractility also varies in response to several factors including temperature, accumulation of lactic acid, fatigue and the overall length of the muscle. Each muscle fibre has an optimal length from which it exerts its maximal contractile force. When it contracts from lengths significantly longer or shorter than the optimum, the force it exerts is correspondingly reduced. The efficiency and force of contraction of the muscle as a whole are also increased by preliminary 'setting', which serves to take up the slack in the non-contractile elements, such as tendons and connective tissues, concerned in the movement.

The activity of every muscle is controlled by signals sent along nerve fibres from the brain and spinal cord. Muscles consist of bundles of muscle fibres. Each motor nerve cell controls a number of muscle fibres (cells), which contract fully and simultaneously whenever the nerve cell sends out a signal. Each group of muscle fibres with its controlling nerve cell is called a *motor unit*. There are only five or six muscle fibres per motor unit in small muscles performing precise movements, such as moving the eye. In large powerful muscles, like those in the buttock or thigh, there may be thousands of muscle fibres in each motor unit. The smallest muscular contraction is the response of a single motor unit and the strongest represents the full activity of all the units.

The isolated contraction of a single large motor unit produces a muscle twitch which exerts a force 1–10 newtons. If the motor unit contracts again before the previous mechanical response is over, the second contraction is added or superimposed upon the first. With high rates of motor-unit firing there is not time for the mechanical response to die away and a sustained contraction or tetanus results (Fig. 5.8).

Normally the motor units contract or fire at rates varying from 1 to about 50 per second. The fine gradations of muscular contraction, between transient muscle twitches and full voluntary contraction, are achieved by increasing the rate of firing of the individual motor units and by recruiting more motor units which in turn fire at faster rates. Sustained muscular contraction results from the rapid asynchronous firing of many motor units. Probably the motor units in a muscle never contract all together except under the most extreme conditions. Normal muscular contractions are maintained by the alternation of different motor units, thus allowing the individual motor units time to recover from fatigue. In everyday life continuous sustained and unvarying contraction of muscle is not a common occurrence. Normal human activity involves intermittent contraction and relaxation of individual muscle fibres, but these muscular contractions are smoothly integrated and coordinated so that the resultant muscular effort is steady and produced with the minimum of exertion.

Neuromuscular transmission

The motor nerve fibres divide into many smaller branches, which run in grooves on the surface of the muscle fibres to *motor end plates*. These are expansions of the nerve fibres which make extensive contact with the muscle fibres but remain anatomically separate from them. They are comparable to the synapses found in the

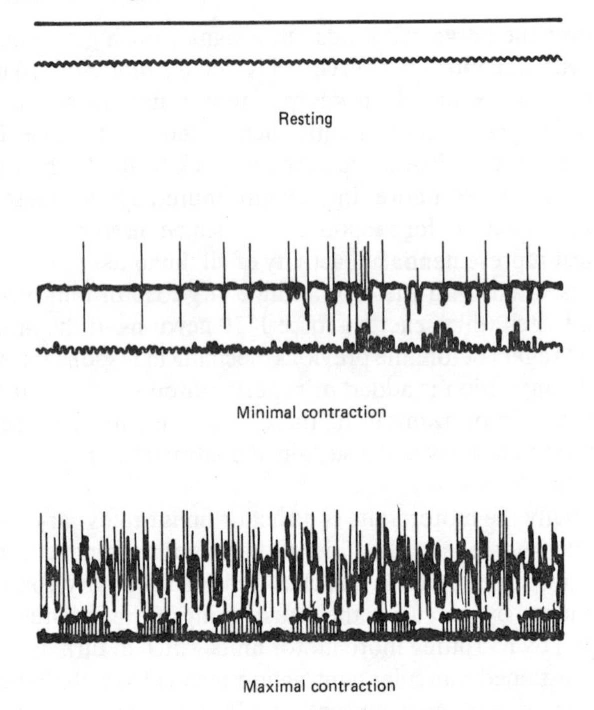

Fig. 5.8 Records of the electrical activity of skeletal muscle. The resting muscle is silent but minimal contraction is associated with the firing of a single motor unit. Stronger contraction results from the recruitment of more motor units and increase in their firing rates. Maximal contraction is associated with an 'interference pattern' of many units firing at high rates

central nervous system (p. 134). In them are numerous vesicles, each containing about 10 000 molecules of acetyl choline. When a nerve impulse reaches the motor end plate the calcium channels are opened and there is a sudden inflow of Ca^{2+} ions, which causes the vesicles to fuse with the terminal membrane and discharge their acetyl choline. The acetylcholine rapidly attaches to receptors on the muscle fibre membrane. Each packet of 10 000 acetyl choline molecules opens up about 2 000 channels through which about

20 000 Na^+ ions pass. This almost abolishes the muscle-fibre membrane potential and sets up an electrical disturbance, or action potential, which spreads down the muscle fibre at a speed of about thirty metres per second. The process is comparable to the transmission of the nerve impulse described on page 136 and is followed by a similar refractory period. The vesicles are reformed and refilled with acetyl choline for the next impulse to utilise.

The action potential leads, after an interval of one or two milliseconds, called the *latent period*, to contraction of the myofibrils inside the muscle fibre. The duration of the resulting shortening depends on the function of the muscle. In small, rapidly acting muscles like the eye muscles it lasts 10 milliseconds (10/1000th sec) or less, but in a powerful calf muscle like the gastrocnemius it may last 30 milliseconds (30/1000th sec) or more.

Each nerve impulse produces only one contraction of the muscle fibre, because the acetyl choline liberated is rapidly hydrolysed by the enzyme cholinesterase, which is abundant at the motor end plates. Sustained muscular contraction results from trains of nerve impulses causing successive contractions of muscle fibres in motor units over a wide area of the muscle (see p. 51). The muscle fibre is incapable of responding to a fresh nerve impulse, with its release of acetyl choline, for a short time after it has contracted. During this period of inexcitability, called the *refractory period*, the muscle protein is recovering its ability to contract and the sarcolemma is repolarising.

6

Diet and Nutrition

The ultimate source of all human energy is the sun acting through the mechanism of photosynthesis described above (p. 2). The green plants on land and in the sea are eventually consumed by animals, which are themselves consumed or broken down after death so that the cycle rolls endlessly on. Other forms of energy are available to us from fossil fuels, wind, water and even nuclear fission but they are not assimilated by the body in the same way as the products of the food cycle.

The type, source and proportion of the carbohydrates, proteins and fats in the human diet vary in different parts of the world. The composition of various foodstuffs differs and most foods are mixtures of the three basic nutrients. Fig. 6.2 shows the composition of the commoner foodstuffs. Humans can exist on almost any type of diet if it provides enough fuel for energy and also adequate amounts of water, protein, minerals and vitamins.

Water

The adult human body contains 60–65 per cent water. It is essential for life, as a transport medium and as a solvent in which body chemistry takes place. Absence of water causes death in days, but starvation can be survived for weeks. Normally the amount of water lost by the body is balanced by the amount of water it acquires. The daily exchange of water is shown in Table 6.1. The formation of water by metabolic process has already been explained (p. 10).

Intake		*Output*	
Drink	1300 ml	Urine	1500 ml
Food	850 ml	Breath	400 ml
Metabolism	350 ml	Skin	500 ml
		Faeces	100 ml
Total	2500 ml	Total	2500 ml

Table 6.1 The daily exchange of water

12.5 millilitres of water are formed in the release of 100 calories by the metabolism of foodstuffs.

Distribution of water

An adult male body weighing 70 kilograms (11 stone) contains 45 litres (10 gallons) of water (Fig. 6.1). 27 litres of this, the intracellular fluid, is inside the cells and the remaining 18 litres is the extracellular fluid, which includes the blood. There are very important differences between the composition of the fluid inside and outside the cells. The extracellular fluid contains sodium, potassium, calcium, phosphorus, chloride, bicarbonate, glucose and urea. The intracellular fluid contains very little sodium and chloride but far more potassium and protein. These differences are maintained within very narrow limits by a constant interchange. The cells are continually taking in oxygen and nutrients from the extracellular fluid and discharging carbon dioxide and waste products into it. Elaborate mechanisms control these exchanges so that the most suitable conditions for cell life and function are preserved.

Minerals

Minerals such as sodium, potassium, calcium, phosphorus, iron, chlorine, sulphur, iodine and minute traces of copper, manganese, zinc, cobalt and fluorine are essential for life. They are plentiful in most normal diets, but calcium (in milk and cheese) and iron (in meat and vegetables) are sometimes lacking. Minerals are essential for maintaining the osmotic pressure (p. 12) of the body fluids, and the normal function of the cells and tissues, especially nerve, muscle and bone.

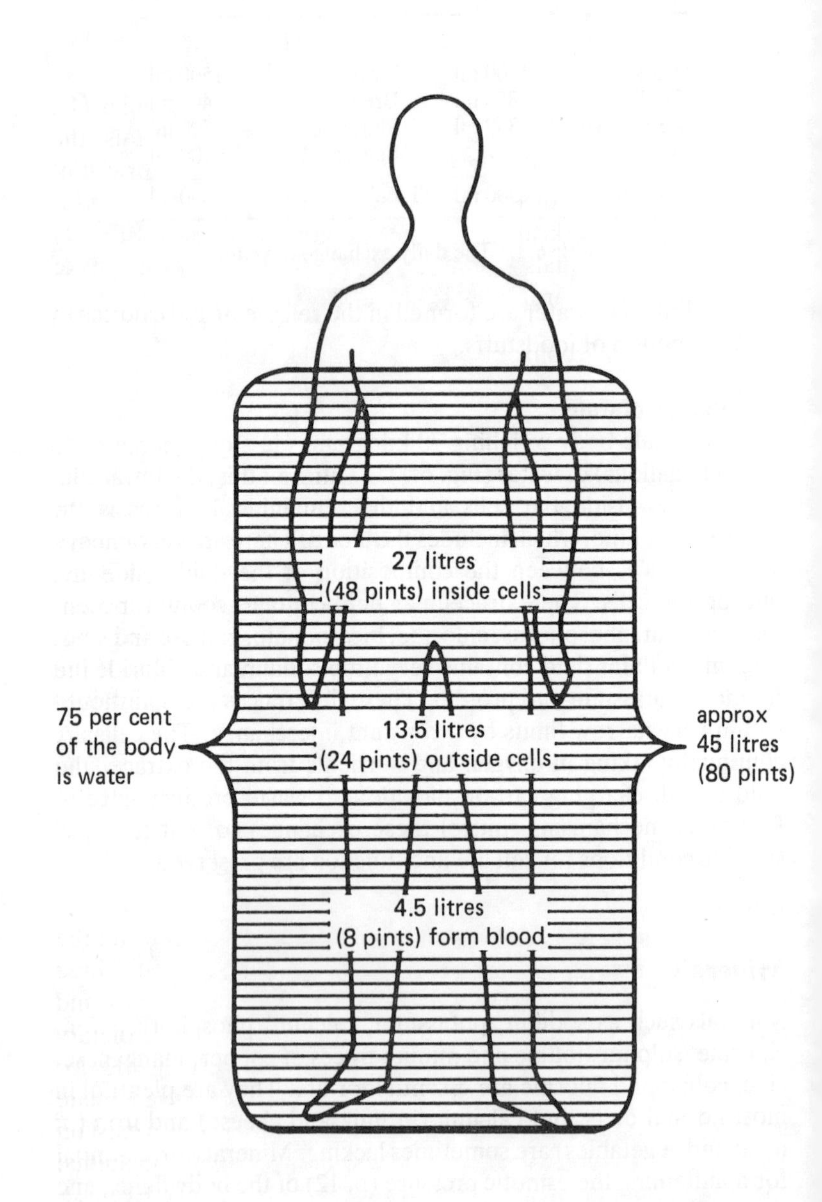

Fig. 6.1 The distribution of water in the human body

Diet

The fuel value of food can be measured in calories or joules (J), which are units of heat. The calorie is the heat required to raise the temperature of 1 g of water from 15 to 16 °C. The calorie unit normally used in medicine and dietetics is 1000 times larger and is called a kilocalorie (kcal) or Calorie. 1 Calorie equals 4.2 joules (J) and 1 kilocalorie equals 4.2 kilojoules (kJ). The heat equivalents of the main foodstuffs are:

Carbohydrate	17 kJ/g (4 kcal/g)
Fat	38 kJ/g (9 kcal/g)
Protein	17 kJ/g (4 kcal/g)

The basal energy requirement of a resting adult is 5000–6300 kJ (1200–1500 kcal) per day, depending on size. A man engaged in light work requires up to 12 600 kJ (3000 kcal) per day and for heavy work 16800–21000 kJ (4000–5000 kcal) will be needed daily. The figures for women are about a fifth less than those for men. Children under the age of 12 utilise correspondingly less energy but children over that age have similar requirements to adults. If the energy used by the body exceeds the calories eaten, weight should be lost, while conversely an excess of fuel will be stored as fat. A deficit of 3780 kJ (900 kcal) daily would be made good by the oxidation of 100 g of fat per day, because fat has a calorie value of 38 kJ (9 kcal) per g.

Foodstuffs

The energy for bodily activities and the raw materials needed for the construction and repair of the body all come from the food. Three main classes of nutrients are available – carbohydrates, fats and proteins. The type, source and proportion of these basic dietary constituents very markedly in different parts of the world for geographical, economic or religious reasons. If there are adequate supplies of water, vitamins and proteins, human beings can exist on the most varied types of diet. Some vegetarians will eat no animal food at all, obtaining their essential proteins from plants. Even so, most people eat a mixed diet and a few live almost entirely on protein and fat.

Carbohydrates

Carbohydrates are compounds of carbon, hydrogen and oxygen. They form the main source of energy in most human diets. Cereals, potatoes or rice are the staple foods and the average diet contains 250–500 g per day. The carbohydrate of plants is mainly cellulose, starch and simpler sugars such as glucose, fructose and sucrose. Cellulose forms the walls of all plant cells and is not digestible by humans. The effect of cooking is to rupture the cellulose coating of the starch granules in plants making the starch accessible to enzyme action. The other carbohydrates are transformed by enzymes into glucose before being absorbed into the body and utilised. Glucose is a white, crystalline solid with a sweet taste and is present in human blood in a concentration of about 5 mmol/l (80 mg per cent).

Glucose not burnt for energy (oxidised) is converted to glycogen and deposited in the muscles and the liver or transformed into fat and stored more permanently. The amount of carbohydrate in the diet can be greatly reduced but at low levels fat metabolism becomes disordered and acids accumulate in the body with serious effects on health.

Fats

Fats are the richest source of energy in the diet. They have a high fuel value, increase the palatability of food and tend to satisfy hunger. They are expensive but some intake of fat is essential for healthy life. The normal intake is 70–100 g per day. They also are compounds of carbon, hydrogen and oxygen but are much more complex than carbohydrates. Each fat molecule consists of one molecule of glycerol – a form of alcohol – and three molecules of fatty acid. Different combinations of fatty acids form chemically different fats. Fat in the food is broken down into glycerol and fatty acids before absorption into the blood. Eventually these constituents are either converted into energy or reformed into fat and stored.

Proteins

Minimal quantities of protein are essential for growth and for the replacement of tissue protein, which is constantly being broken down during normal life. The estimated ideal amount of protein is 1 g per kg body weight, i.e. 70 g per day for an average man.

For economic reasons the amount of protein taken is often less and protein undernutrition is one of the commonest forms of deprivation.

Proteins are very large, complex molecules formed by the intricate linkage of thousands of molecules of amino acids. In the body there are twenty-three different amino acids, all composed of carbon, hydrogen, oxygen and nitrogen. All but ten of them can be formed inside the body. These ten essential amino acids, which must form part of a healthy diet, are plentiful in animal foods but some are scarce in plants. The large protein molecules are broken down and absorbed as amino acids, which are then either relinked to form necessary proteins or used for energy. Proteins not only provide most of the structural material of the cell but also control the multitude of chemical reactions taking place simultaneously in each cell. They are essential for both the survival and the special functions of every cell.

Animal proteins, which are the most expensive, contain the essential amino acids in adequate quantitites and proportions. The essential amino acids are much scarcer in vegetable proteins but beans, nuts and potatoes are a better source of protein than grains and root vegetables. It is noteworthy that all amino acids stimulate metabolism and increase the rate of production of heat in the body by approximately 20%. This stimulant effect is called the specific dynamic action of protein and it is utilised in the design of diets intended for weight reduction, which usually contain large amounts of protein.

Composition of the diet

As stated above, an average diet contains 70–100 g protein, 70–100 g fat and 400–500 g carbohydrate with a total energy value of 10 500–14 000 kJ (2500–3300 kcal). The actual fat, protein and carbohydrate content of the raw foods is much less than their total weight because of their high water content. For example, the average composition of various foods is as shown in Table 6.2 and Fig. 6.2. In addition most foods contain cellulose, collagen fibres and other indigestible material which provide no energy, although they are valuable in adding bulk to the contents of the intestine. It follows that the weight of the material in a diet must be considerably greater than the weight of the active constituents listed above. In

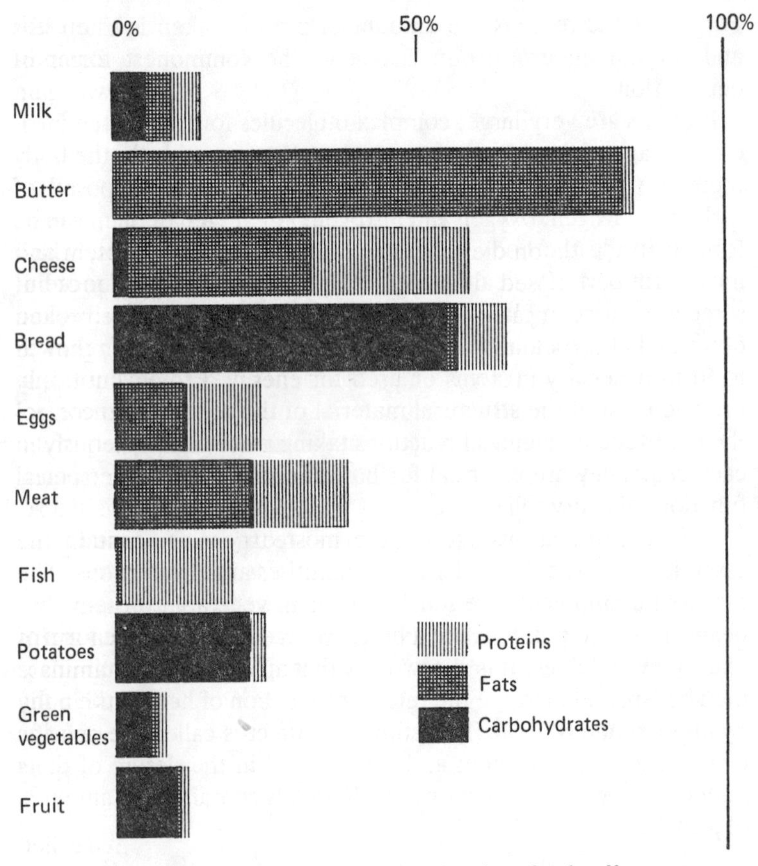

Fig. 6.2 The composition of various foodstuffs

	Protein (%)	*Fat* (%)	*Carbohydrate* (%)	*Water* (%)
Lean meat	30	8	0	60
Eggs	13	10	0	75
Milk	3	4	4	88
Bread	9	1	52	35

Table 6.2 The average composition of various foods

practice appetite and custom regulate the intake of food fairly well. Regulation tends to lag with increasing age when energy expenditure becomes reduced and many middle-aged people will gain weight unless they consciously reduce their food intake.

Vitamins

If animals are fed on diets adequate for the supply of protein and energy but composed of pure, artificial foodstuffs, i.e. protein, carbohydrate and fat, it is found that they do not thrive and eventually die. Additional factors which are essential for growth and health are found in very small quantities in natural, but not in synthetic foodstuffs. Before their structure was known these accessory food factors or vitamins were designated by the letters of the alphabet. They can be divided into two groups according to their solubility in fats or water. The fat-soluble vitamins are labelled A, D, E and K, and the water-soluble ones are the B group and C.

Fat-soluble vitamins

These are probably absorbed from the small intestine in the form of finely emulsified fat globules. Deficiency of fat-soluble vitamins is therefore likely to occur in disease of the small intestine and in the absence of bile salts, which are essential for full absorption of fat as well as from dietary inadequacy.

VITAMIN A

This is a complex alcohol derived from plant pigments called carotenes, which are transformed into vitamin A in the liver. The precursors of vitamin A can be obtained from vegetables and fruit and the vitamin itself is present in milk fat, liver and animal fats. The normal requirement of vitamin A is 5000 international units (IU) per day. Vitamin A is essential for the integrity of epithelial cells and in its absence the epithelium of the eye, the lung and the genito-urinary tract becomes dry and horny and liable to infection. Vitamin A is also essential for the formation of visual purple in the eye and deficiency may cause night blindness (p. 170).

VITAMIN D

Vitamin D prevents rickets in children and animals. It regulates the absorption of calcium from the intestine and thus contributes to the normal calcification of bone. Vitamin D is really a group of compounds which are the precursors of a metabolically active hormone found in the body. These precursors are stored in human skin, where the ultraviolet light from the sun converts them into a substance which passes to the liver and then to the kidney to be transformed into 1,25-dihydroxycholecalciferol. This circulates in the blood and increases the absorption of calcium from the intestine. No additional vitamin D is needed in the tropics but elsewhere it is an essential part of the diet and is found in eggs, butter, milk and fish liver.

The normal daily requirements are 400 international units for a child and 800 IU for an adult. Toxic effects, including calcification of the lungs and kidneys, may occur if very large doses of, say, 100 000 IU are given for many months. Deficiency of vitamin D in children causes rickets, in which there is a failure to deposit calcium and phosphorus in the growing bones with resulting bending, deformity and loss of stature. A similar condition occurring in adults after growth has stopped causes much pain but less bony deformity and is called osteomalacia. Calcium and phosphorus are not laid down to replace the normal removal of bone which constantly occurs (p. 36) so that spontaneous fractures and bony deformity result.

VITAMIN E

These compounds are called tocopherols. Their function in man is uncertain but in animals their lack causes sterility and muscle wasting. They are found in milk, eggs, muscle and especially wheat germ oil.

VITAMIN K

Vitamin K, which is one of a group of compounds called naphthoquinones, is found in the leaves of green vegetables. It is used by the liver in the formation of prothrombin (p. 84), which is essential for the normal process of blood coagulation. In patients with obstructive jaundice, defective vitamin K absorption leads eventually to low prothrombin levels in the blood, but this can be corrected by the injection of vitamin K subcutaneously. In liver damage the prothrombin level is also low, but in this case the level cannot be returned to normal by the injection of vitamin K because the liver cannot synthesise prothrombin satisfactorily.

Water-soluble vitamins

VITAMIN C

Ascorbic acid – vitamin C – occurs in citrus fruits, berries, melons, tomatoes, green vegetables and potatoes. It is easily destroyed by cooking. Its absence from the diet for more than three months leads to scurvy, in which the structure of bone, dentine, cartilage and connective tissue becomes abnormal. As a result, haemorrhages, loss of teeth, spontaneous fractures and poor wound healing are found. Scurvy caused great loss of life during early sea voyages. The suprarenal glands contain large amounts of vitamin C which is rapidly used up in infection and fever. This has led to the belief that the vitamin may be related in some way to the reaction of the body in stress. 100 mg per day is the normal requirement.

VITAMIN B COMPLEX

It was originally thought that deficiency of vitamin B led only to beri-beri in tropical countries. It is now known that vitamin B consists of a whole series of unrelated compounds which have many activities. Only a few of them can be mentioned here.

THIAMINE

Thiamine, which is found in cereals and meat, acts as an enzyme in the later stages of carbohydrate utilisation. Its lack leads to beri-beri, in which cardiac failure, oedema and neuritis occur. 1–2 mg per day is the normal requirement.

RIBOFLAVINE

This vitamin is found in meat and milk. Its absence leads to changes in the eyes, the skin of the face and mouth and the lips and tongue. The conjunctiva becomes inflamed and the eyelids are red and swollen. The skin of the face is rough, cracked and scaling while the lips swell and the angles of the mouth become cracked and sore. The tongue turns bright red and is both tender and painful. 1–2 mg per day is the normal requirement.

NICOTINIC ACID

This is found in yeast and meat. Its lack causes pellagra, in which diarrhoea, mental changes and skin disease occur. 15 mg per day are normally required.

FOLIC ACID GROUP

Folic acid is present in green leaves, liver and kidney. It is essential for normal blood formation. 50 μg daily are required.

CYANOCOBALAMIN – VITAMIN B_{12}

Vitamin B_{12} is found in animal products and liver. It can only be absorbed from the lower end of the small intestine in the presence of 'intrinsic factor' which is secreted by the mucosal lining of the stomach. Vitamin B_{12} deficiency usually arises from defects in the stomach or intestine. It sometimes occurs in very strict vegetarians. It is essential for the normal maturation of the red blood cells and lack of it leads to a special kind of blood disorder – pernicious anaemia (p. 82). This is sometimes accompanied by degenerative changes in the spinal cord and peripheral nerves, causing sensory changes and muscular weakness. 2.5 μg are needed daily.

Basal metabolism

The total heat production is a measure of the total metabolic activity and energy expenditure of the body. It is the sum of the energy used for the maintenance of vital processes (the basal metabolism), and that used in doing work. The basal part can be estimated in the resting, fasting individual from the oxygen used over a period. It is normally about 6300 kJ (1500 kcal) a day or 273 kJ (65 kcal) an hour. The metabolic rate can increase to 4200 kJ (1000 kcal) an hour in violent exertion. This energy output is normally balanced by heat loss but after severe exercise the rectal temperature may rise by 2–3 °C for up to half an hour.

Regulation of body temperature

Most body heat arises from muscular exercise but active organs like the liver, endocrine glands and skin also participate. Liberation of adrenaline and thyroxine into the circulation also increases the rate of metabolic activity, and heat production, in the tissues.

The nude body loses 50–60 per cent of its heat by radiation from the surface, 15 per cent by conduction or transfer into the surrounding air (convection) and the rest by evaporation. These mechanisms are only effective if the external temperature is lower than that of the body; if it is higher the body becomes hotter. Conduction–convection dissipates heat because the air molecules nearest the body become hotter and more agitated. They tend to move away from the surface of the skin and are then replaced by cooler air molecules; these movements are called convection currents. If the air next to the body is kept still by clothing the heat-losing mechanisms become less effective. This is the reason for the warmth engendered by cellular blankets, eiderdowns and string vests. Evaporation of 2 g of water removes 4.2 kJ (1 kcal) of heat and the body can produce as much as 2 litres of sweat per hour, which dissipates 4200 kJ (1000 kcal). Sweat is a watery fluid containing about 0.3 g per cent salt (NaCl) and small amounts of other electrolytes and some waste products. Excessive sweating can cause serious loss of water and salt. If the humidity of the air is high, evaporation is reduced, with consequent diminution of heat loss causing discomfort and even danger in very hot, moist conditions. Both conduction and evaporation are greater in moving air and so fans make buildings more tolerable in hot weather.

Clothing interferes with heat loss by all three mechanisms, largely because of the insulating properties of the air trapped in its meshes. Wet clothing loses this power of insulation and heat loss is rapid. In the Arctic wet clothes are therefore dangerous but in the tropics they are beneficial because evaporation occurs from their surface.

The temperature of the body depends upon a balance between heat production and heat loss. The control of this balance is effected by the nervous system in response to nerve impulses which originate in the *hypothalamus* (p. 143). The hypothalamus is sensitive to the temperature of the blood flowing through it and acts as a regulator or thermostat. It behaves as if set at 36.6 °C and if the blood temperature falls below this level the arterioles in the skin constrict, diminishing the flow of blood through it and therefore reducing its temperature. The loss of heat from the body depends on the difference between the temperature of the skin and that of the surrounding air. In cold conditions the skin becomes pale, cold and relatively bloodless. Heat loss from the body is greatly reduced because the mechanisms described above are inoperative and because the superficial body fat insulates the deeper layers. The body has little control over its heat production apart from varying the amount of muscular work done, e.g. by resting when hot and taking vigorous exercise when cold. Adrenaline is released into the circulation increasing the rate of metabolism and, if necessary, shivering occurs causing much heat production by muscular contraction. In animals the hair stands out and entraps more air to act as an insulating cover to the body.

If the temperature of the blood rises, the hypothalamus initiates sweating and reduces sympathetic nervous activity. The rate of blood flow through the skin increases greatly because the superficial blood vessels dilate and the rate and output of the heart are augmented. The increased skin temperature leads to greater heat loss as does the excretion of water in the expired air of panting animals.

7

The Alimentary System

The materials needed for energy, maintenance and repair are extracted from the food and absorbed from the alimentary tract. This is a long coiled tube which runs from the mouth to the anus. It is 3.5–5 cm (1.5–2.0 inches) in diameter throughout most of its length but there are two wider portions, the stomach and the colon, where the movement of the contents is slowed down. At each end rings of muscle seal off the tube, except when material is entering or leaving. It is lined by mucous membrane arranged in folds to give a large surface area which enables it to absorb with great efficiency the various chemical substances produced by digestion. At the same time 5.5 litres (10 pints) or more of water are absorbed daily.

The alimentary tract

Structure of the alimentary tract (Fig. 7.1)
The *oesophagus* or gullet is a muscular tube about 25 cm (10 inches) long connecting the *pharynx* to the *stomach*. It passes down in front of the vertebral column in the chest and passes through the diaphragm at the level of the 10th thoracic vertebra. It lies behind the left lobe of the liver and enters the cardiac end of the stomach.

It is made up of four layers. The outer, fibrous layer is composed of dense connective tissue with many elastic fibres. This surrounds the muscular layer, which forms a continuous outer tube of longitudinal fibres with thinner rings of circular fibres at the upper and lower ends. There is then a *submucosa* containing blood vessels and nerves underlying the *mucosa*. This is thick with distensible longi-

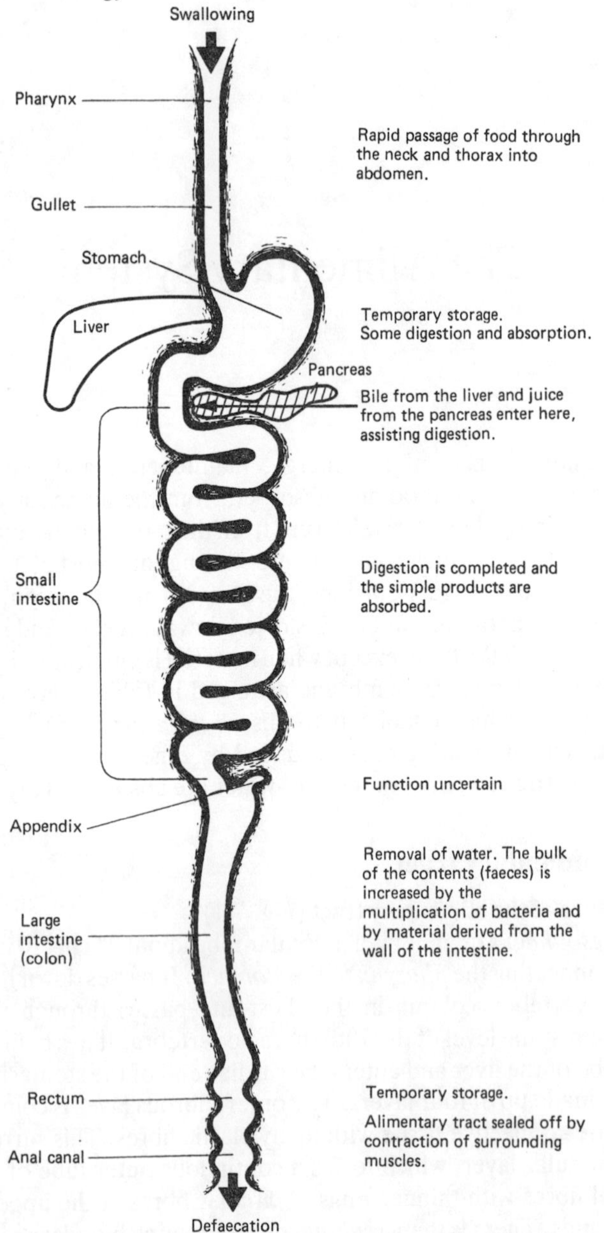

Fig. 7.1 The overall arrangement of the digestive tract

tudinal folds which are covered by tough squamous epithelium.

The *stomach* is the widest part of the alimentary tract. It is usually represented as lying in the upper and left parts of the abdominal cavity, but there is much variation related to posture and the state of digestion. At times its lower border may lie in the pelvis. It consists of a dome-shaped *fundus*, usually filled with air and the main body or digestive portion. This narrows down into the *pylorus*, which leads to the duodenum. It is covered by a smooth serous coat – the *peritoneum* – covering a thick muscular layer. This has longitudinal, oblique and circular fibres. The internal mucous membrane is a thick, heavily folded lining containing the profusion of gastric glands with several different types of cell. The chief or peptic cells are cuboidal with plentiful rough endoplasmic reticulum. They produce the digestive enzymes. There are several other cell types present including mucus-secreting cells. The functions of the others are less certain.

The pyloric end of the stomach opens into the *small intestine*, which is about 5–6 m (16–20 feet) long and ends at the *large intestine* in the right lower abdomen. Its first part, the *duodenum*, is firmly attached to the posterior wall of the abdomen and is about 25 cm (10 inches) long. The pancreatic and common bile ducts enter the descending part of the duodenum through a papilla.

The first 2 m (6 feet) of small intestine is called the *jejunum*, the rest being the *ileum*. The jejunum is thicker, wider and more vascular than the ileum. Its mucous membrane is arranged in large crescentic folds projecting inwards and covered by enormous numbers of minute finger-like villi. These folds are less obvious in the ileum and are absent at its lower end. The wall of the intestine has the same four layers as the stomach although the secretory cells have different functions. The whole small intestine is arranged in coils suspended from the posterior abdominal wall by a mesentery containing blood vessels, lymphatics and nerves. In the ileum especially there are numerous collections of lymphatic follicles (Peyer's patches) varying from 2 to 10 cm (1–4 inches) in length. In general digestion continues in the jejunum while absorption occurs mainly in the ileum.

The *colon* is about 1.5 m (5 feet) long and extends from the ileum on the right up and around the abdominal cavity to end at the anus. It has a layer of peritoneum surrounding muscle. The muscle is arranged in three longitudinal bands, which are shorter than the inner circular layer of muscle fibres so that the colon is puckered transversely into haustrations. There is then a submucosa supporting the mucous membrane, which is pale and smooth. It has numerous mucus-secreting glands but no villi.

Just below the junction of the ileum and the colon there is a narrow worm-like tube 2–20 cm (1–8 inches) long – the *vermiform appendix*. Its structure is similar to the small intestine but its function is not known.

The colon ends in the *rectum*, which lies in the pelvis and terminates at the anal canal. It is about 12 cm (5 inches) long and can be greatly distended. It has a thick muscular coat, which merges below at the ano-rectal junction with the powerful internal anal sphincter. Outside this, and more superficially, is the external anal sphincter, which is under voluntary nervous control.

Function of the alimentary tract

In the mouth the food is broken down by the teeth and mixed with saliva produced by three pairs of salivary glands – parotid, submandibular and sublingual – situated around the lower jaw. The thought, smell or sight of food will produce saliva by means of a conditioned reflex, but the presence of food, a pebble or acid in the mouth causes salivation by simple reflex action (p. 152). Normally 1–1.5 litres of saliva are secreted daily. It is a colourless, opalescent, sticky and slightly acid fluid which contains mucin for lubrication and an enzyme called *ptyalin* (salivary amylase), which converts starch into maltose. The secretion of saliva is adapted to the type of food which is taken. Meat, which is swallowed without initial digestion, excites a strong flow of lubricant mucin, but dry biscuits lead to the secretion of much serous saliva containing amylase.

The chief functions of saliva are physiological and protective; the breakdown of starch into maltose by amylase is transient and of little practical importance. Saliva moistens and lubricates the food, helping the formation of food masses suitable for swallowing. The moistened food is pushed back by the tongue into the oesophagus. The first stage of swallowing is voluntary. The food mass, or bolus, is pushed upwards and backwards into the pharynx by the tongue. The soft palate moves upwards and closes the back of the nose. The larynx moves upwards towards the epiglottis, shutting off the

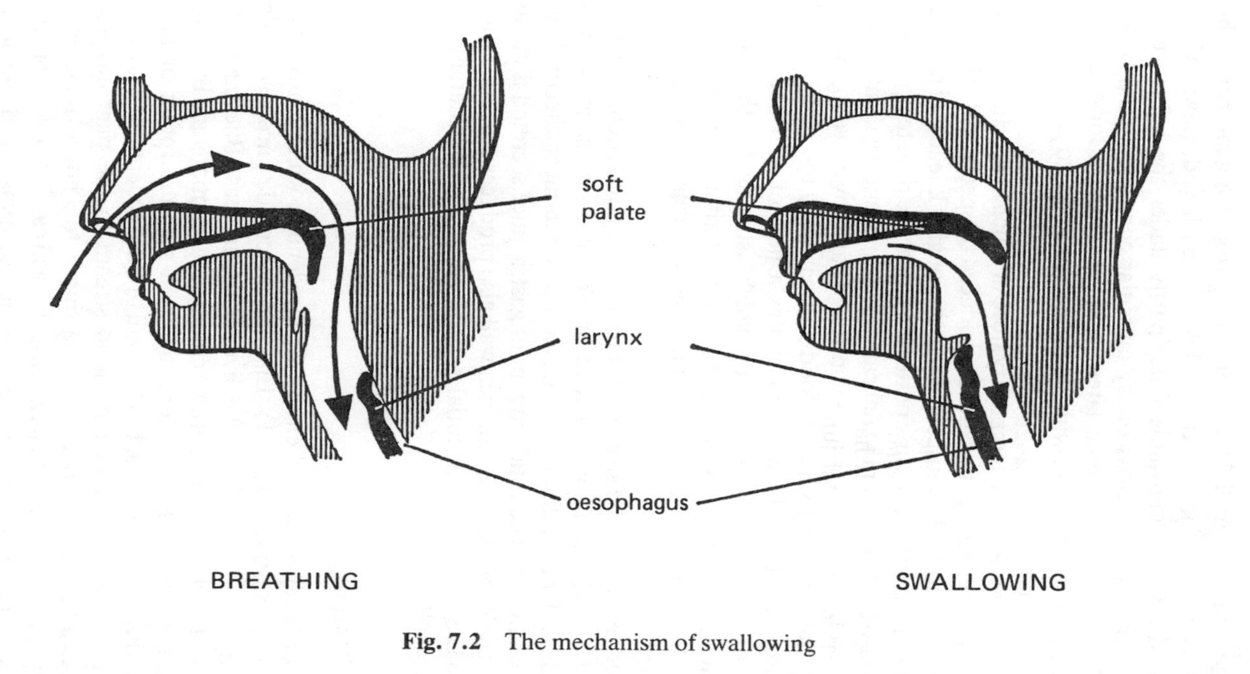

Fig. 7.2 The mechanism of swallowing

laryngeal cavity (Fig. 7.2), and breathing stops for a moment. The oesophagus now forms a tube, closed at each end, down which liquids drop and semisolids slide, partly under the influence of gravity and partly moved on by muscular contraction and relaxation. Two or three seconds later the sphincter at the lower end of the oesophagus opens allowing fluids and semisolids to pass through into the stomach.

The stomach acts as a temporary store in which hydrochloric acid, mucin and the enzyme pepsin are produced by specialised cells and added to the food. Its muscular walls thoroughly knead and mix the food until it is a thick creamy paste (chyme) at the right temperature and consistency to enter the delicate duodenum and small intestine. The secretion of gastric juice is controlled partly by nerve signals set off by the sight and smell of food and partly by the type of food entering the stomach. Meat promotes a flow of acid because it liberates a hormone – gastrin – which stimulates the acid-producing cells in the wall of the stomach. Irritants evoke a large amount of mucin, which lubricates and protects the wall of the stomach.

Food remains in the stomach for four or five hours, depending on its nature. The proteins begin to break down into simpler compounds under the influence of pepsin, which acts most effectively in acid conditions. The total amount of gastric juice secreted is about 3 litres in 24 hours. The stomach wall also produces *intrinsic factor* (pp. 64, 82) which is essential for the absorption of vitamin B_{12} in the small intestine.

Intestinal digestion

The partly digested food enters the duodenum, which leads on to the small intestine (Fig. 7.1). Bile and pancreatic juice also drain into the duodenum from the liver and pancreas. Bile pigments produce the characteristic brown colour of the motions and bile salts are important for the efficient digestion and absorption of fats. When the acid chyme reaches the wall of the duodenum a hormone called *secretin* passes in the blood stream to the pancreas and induces a flow of pancreatic juice lasting 2. to 3 hours. Pancreatic juice is alkaline enough to neutralise the acid carried down from the stomach and contains the enzymes lipase, trypsin and amylase,

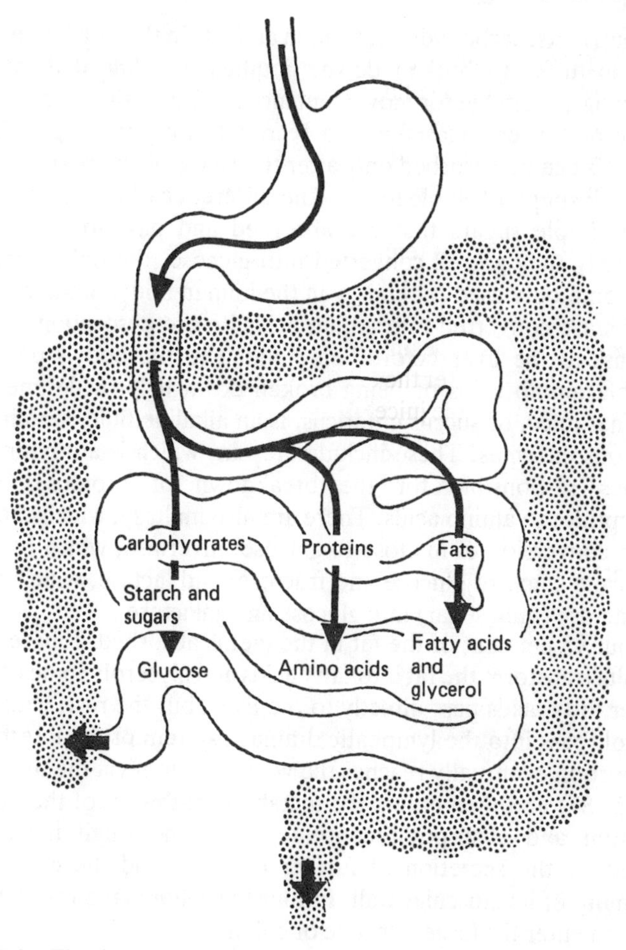

Fig. 7.3 The digestion and absorption of foodstuffs in the small intestine

which help to digest the fats, protein and starch of the diet into fatty acids, glycerol, amino acids and glucose (Fig. 7.3).

The intestinal contents are thoroughly mixed and passed along the gut by muscular movements. Ripples of contraction arise about eighteen times per minute at the duodenal end of the small intestine and travel rapidly down towards the junction of the ileum and colon.

Different types of food have greatly varying contents of the three basic nutrients, carbohydrate, protein and fat. In the small intestine the foodstuffs are broken down, liquified and kneaded into a creamy paste suitable for movement along the intestine and for the activity of the enzymes that are secreted along its length. The foodstuffs can be absorbed only after they have been broken down into much simpler soluble forms. The different carbohydrates give rise to simple sugars that are absorbed and pass to the liver. Eventually they are all converted into glucose, the only form of sugar commonly used for energy in the human body. Similarly the proteins are converted into amino acids before absorption. They also pass to the liver before being added to the body's pool of available amino acids or being broken down to provide energy. Intestinal juice, or succus entericus, is an alkaline fluid containing numerous enzymes. These include erepsin, which is a mixture of enzymes responsible for the breakdown of proteoses and polypeptides to amino acids. There are also maltase, which catalyses the conversion of maltose to glucose; invertase, which breaks down cane sugar to glucose and fructose; and lactase, which converts lactose (milk sugar) into glucose and galactose.

About 95 per cent of the fat in the diet is absorbed. Lipase and bile salts transform the fat into fatty acids and glycerol. Some of the simpler fatty acids pass directly to the liver, but the rest, with the glycerol, pass into the lymphatic drainage system of the intestine. This portion eventually reaches the venous system via the thoracic duct (p. 87) and then enters the metabolic pathways of the body. Digestion and absorption continue along the small intestine, assisted by the secretion of further enzymes and the continual movement of its muscular wall. In four to six hours the residues of digestion enter the large intestine or colon.

The colon

The colon has a muscular wall with a lining of mucous membrane through which large amounts of fluid can readily pass. When food leaves the stomach there is a reflex discharge of intestinal contents from the lower ileum into the colon through the ileo-caecal sphincter – the gastro-colic reflex. There is also a passive movement of material from the ileum to the colon about four or five hours after taking food. Normally about 350 to 500 g of faecal matter enter the colon daily, water is absorbed and the faeces become formed in the next 24 to 36 hours. There is a daily deposit of 80 to 200 g of faeces, consisting of bacteria, cellulose from the food and cells shed by the intestinal wall. The brown colour of the stools results from the presence of stercobilin, which is formed from the bile.

The bulk of the stools and the regularity of the motions is largely dependent on the amount of residue from the diet. In the last fifty years the food of Europeans and North Americans has become increasingly refined and concentrated. In particular the fibre content or roughage of natural untreated foods has been largely eliminated from the conventional diet. The cellulose fibres from plant cells have no energy value in humans because they are not broken down and assimilated. The husk of wheat has been removed from white bread and other dietary fibre has diminished. It is now realised that such restriction is potentially harmful and liable to produce intestinal disorders such as constipation, inflammation and new growth causing tumours. A normal diet should contain ample roughage as in wholemeal bread, cereals, fruit and vegetables.

The whole process of digestion and absorption is precisely controlled and ordered from mouth to anus. Each part of the digestive system reacts to the amount and nature of the material reaching it but is under the more general control of nearby sections through chemical messengers (hormones) carried in the blood. For example, the amount and type of pancreatic secretion is controlled by two hormones, gastrin and secretin, in proportion to the amounts of acid and protein in the chyme leaving the stomach. The two hormones travel to the pancreas and release precisely the amount and type of pancreatic juice that is needed in the duodenum. Finally, the overall secretory and muscular activity of the whole alimentary tract is coordinated and controlled by signals in a network of nerve fibres running out from the brain and spinal cord – the autonomic nervous system. Impulses in sympathetic nerve fibres reduce, and in parasympathetic nerve fibres increase, the activity of the wall of the intestine.

Defaecation

The passage of faeces from the anus is controlled by two sphincters. There is an internal one consisting of smooth muscle under the control of the autonomic nervous system while the external sphincter of striped muscle is under voluntary control. Defaecation is initiated by reflex stimulation from the lower colon and especially from the rectum. The rectum is normally empty but fills, reflexly, after a meal and there is then a desire to defaecate which can be opposed by voluntary contraction of the external sphincter. When defaecation occurs both the sphincters relax, the abdominal muscles contract, the diaphragm is fixed in the position of forced inspiration and high intra-abdominal pressures develop. A wave of contraction empties the left-hand half of the colon into the rectum and the levator ani muscle pulls the anus up over the faeces. Defaecation usually occurs one to three times daily but variations from these figures are not unusual.

The liver

The liver is the largest gland in the body. It weighs about 1.5 kg (3.3 lb) in the male and 1.3 kg (3.0 lb) in the female. It consists of a large right lobe, which fills the upper right segment of the abdomen from the rib margin almost to the nipple level, and a smaller left lobe, which extends across the mid-line to the level of the other nipple. It is wedge-shaped, reddish-brown in colour and firm to the touch but easily damaged by pressure. It is extremely vascular and bleeds readily after injury. Its upper surface is closely applied to the diaphragm and its under-surface is in close contact with the stomach, colon, kidneys and spleen. The dark blue gall bladder lies under the right lobe and gives off the cystic duct, which joins the hepatic duct to form the common bile duct. This is about 7.5 cm long and 6 cm in diameter. It enters the descending part of the duodenum.

The liver is largely covered by peritoneum. It consists of very large numbers of polyhedral lobules each about 1 mm in diameter. They are formed by layers of cells (hepatocytes) arranged in sheets one cell thick. They form a labyrinthine system of channels and spaces communicating with each other. The blood and drainage systems of the liver ramify in these channels (Fig. 7.4). The hepatocytes have a double blood supply. They need abundant oxygen and this is brought by branches of the hepatic artery. All the blood containing the products of digestion absorbed from the intestine

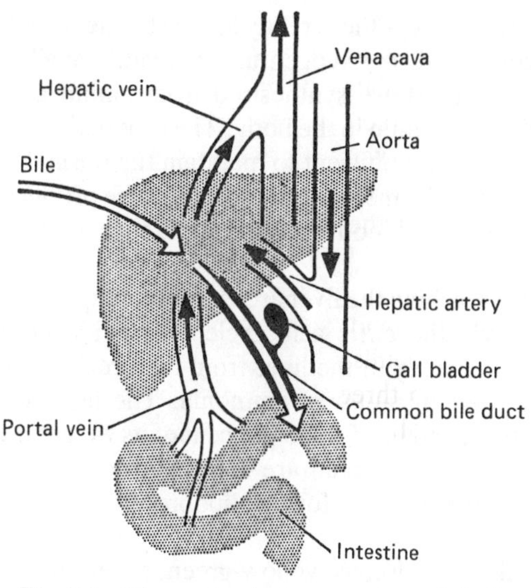

Fig. 7.4 The anatomical connections of the liver

passes along the portal vein. The cells perform their manifold metabolic functions on this blood, which then leaves the liver by the hepatic vein and enters the general circulation through the lower vena cava. Blood in the hepatic vein carries all the varied chemical substances which the liver supplies to the rest of the body. The waste products of liver activity are removed by a separate system of channels – the hepatic ducts – which join together to form the common bile duct. The gall bladder stores and concentrates the bile. Its function is largely mechanical and it is not essential to life. Its muscular coat contracts reflexly when food enters the stomach and bile passes along the bile duct into the duodenum.

Liver function

The hepatocytes have high concentrations of ribosomes, mitochondria, lysosomes and a well-developed Golgi apparatus, reflecting their high level of metabolic activity. They contain many thousands of enzymes needed for the multifarious biochemical processes occurring there. All the nutrients, minerals and vitamins A, D and B_{12}, absorbed from the intestine are stored in the liver or converted into other substances the body needs. Reserve supplies of fatty acids and amino acids are kept there ready for use by the rest of the body. Glucose is converted to glycogen and is readily available on demand. Many proteins are synthesised from amino acids and, in particular, all the albumin in the body. The production rate is about 10 g per day, which is sufficient to maintain the normal level of 5 g per kg body weight. Numerous other proteins are produced, including antibodies and the blood clotting factors such as fibrinogen and prothrombin.

Amino acids reaching the liver are either built up into proteins or broken down via the citric acid cycle to provide energy. Urea synthesis occurs mainly in the liver from the breakdown of amino acids and the detoxification of ammonia. The liver also actively synthesises lipids, and fats form 5 per cent of its weight. Most of the fatty acids entering the organ are converted to triglycerides and enter the blood stream in the form of lipoproteins but some become cholesterol.

Bile is an alkaline, viscous, yellow-green, bitter fluid (pH 8–8.6) which is secreted continuously by the liver at a rate of 0.5–1 litre per day. It is stored and concentrated in the gall bladder, which contracts when fat enters the duodenum and forces bile into the intestine along with the pancreatic juice. The mechanism for this is a hormone called *cholecystokinin*, which is secreted by the duodenal wall when fat enters it. Bile pigments are the waste products of haemoglobin breakdown and are excreted by the liver in the bile after being converted from a fat-soluble to a water-soluble form. Bile also contains bile salts, cholesterol, minerals and, on occasions, drugs or poisons taken into the body. Up to 500 mg of bile salts are synthesised daily from cholesterol and are essential for the digestion and absorption of fat in the intestine through their emulsifying action by the reduction of surface tension.

The liver has an important role in protecting the body from harmful or poisonous substances absorbed from the intestine or circulating in the blood stream. In some cases the toxic substance is joined up to some other chemical and the resulting compound is excreted through the liver, e.g. phenols combine with sulphates to produce less harmful ethereal sulphates which are excreted. In other cases compounds are formed with amino acids or with acetic acid and excreted. The liver metabolises many drugs used in medicine and renders harmless (detoxicates) numerous bacterial and vegetable poisons, including alcohol, which are absorbed from the intestine.

The pancreas

The pancreas is a soft, greyish-pink, lobulated gland, 12–15 cm (5–6 inches) long, stretching across the posterior abdominal wall from the duodenum to the spleen. The pancreatic duct extends from the tail of the gland, near the spleen, to its head, which lies in the curve of the duodenum. The duct joins the common bile duct and the two enter the descending part of the duodenum through a papilla. The pancreas is composed of two entirely different types of glandular tissue. The major part consists of lobulated glandular tissue with very granular cells. These granules contain the pancreatic enzymes in bound or inactive form. They are described on page 72. The endocrine portion of the pancreas consists of about one million Islets of Langerhans randomly distributed in the mass of the organ. These consist of cells with different staining properties. Four-fifths of them are B cells, which secrete the hormone insulin – a major factor in the control of carbohydrate metabolism (pp. 74, 185). The remaining A cells produce glucagon, which is also involved with carbohydrate utilisation.

8

Blood

Blood forms 5–10 per cent of the total body weight and there are more than 5.5 litres (10 pints) of it in an adult. If the blood volume falls after bleeding, thirst or excessive loss of fluid from vomiting or diarrhoea, water moves from the tissues through the walls of the capillaries into the blood stream. This restores the blood volume but dilutes the blood.

Composition

Blood is a tissue consisting of *corpuscles* suspended in a fluid *plasma* in the proportion of 45 parts by volume of corpuscles to 55 parts of plasma. Every cubic millimetre (mm^3) of blood contains over five million coin-shaped corpuscles full of iron-containing haemoglobin. This is a protein which can combine temporarily with oxygen and release it where it is needed. Each cubic millimetre of blood also contains 5–10 000 white corpuscles, which are part of the body's defences against infection. There are also about 250 000 platelets per mm^3 of blood; these are essential for sealing off small defects in the walls of the blood vessels produced by injury or disease and for producing the enzyme that starts off the clotting of blood. If blood plasma or whole blood is allowed to stand outside the body, clotting occurs and a protein (fibrin) is deposited leaving a clear fluid – *serum* (Fig. 8.1). Plasma is, in fact, serum plus 0.3–0.5 g per 100 ml of fibrinogen, the substance from which fibrin is formed. It contains much protein as well as the various gases, nutrients, minerals and waste products.

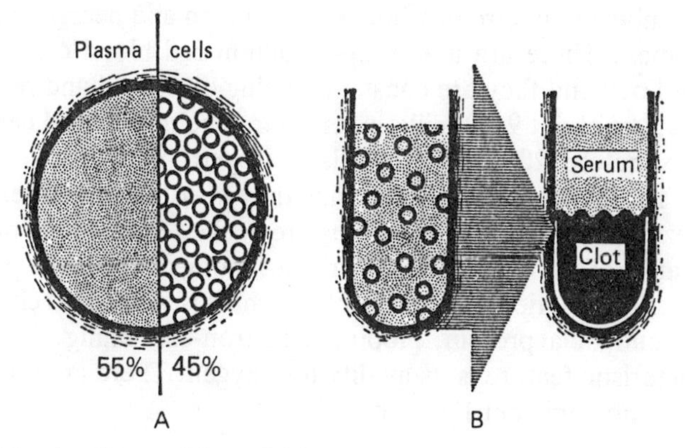

Fig. 8.1 A Composition of blood. **B** Formation of blood clot and serum

The composition of blood plasma is shown in Table 8.1. There are three main proteins – albumin, globulin and fibrinogen. The albumin is essential for the maintenance of osmotic pressure (p. 12) and forms part of the body's reserves of protein. The globulin is concerned with the defence mechanisms of the body and the fibrinogen is essential in the formation of blood clots. The plasma also contains phosphates, sulphates, lactic acid, amino-acids and fatty acids.

Substance	*mmol/1*
Sodium	135–145
Potassium	3.5–6
Calcium	2.5
Bicarbonate	25–30
Glucose	3.5–6.0
Urea	2.5–7.0
Cholesterol	3.5–8.0
Uric acid	0.15–0.45

Table 8.1 The composition of normal human plasma

Red blood corpuscles (erythrocytes)

The red blood cells are discs about 7.5 thousandths of a millimetre (7.4 μm) in diameter and there are about 5.5 million of them in every cubic millimetre of whole blood in men and nearly 5 million in women. There are about 25–30 billion red blood cells in the whole body and they are constantly being destroyed and replaced at a rate of about 9–10 million per hour. The normal red cell lasts approximately 120 days before it is destroyed.

The red blood cells are flexible and they withstand much bending, squeezing and deformation as they are pushed through the narrow capillaries. They consist of an intricate, spongy network of protein filled with a solution of haemoglobin, which is a coloured chemical compound of a protein, globin, with iron-containing haem. Its characteristic feature is its avidity for oxygen. There are normally 15 g haemoglobin per 100 ml blood.

Source and development of red cells

After birth all the red cells are formed from the endothelial cells lining blood-filled spaces (sinusoids) in the red bone marrow which is found in the vertebrae, ribs, pelvis and skull. The process is controlled by a protein – *erythropoietin* – produced by the kidney. Any reduction in the number of red corpuscles or a shortage of oxygen stimulates the marrow to make replacements, but various raw materials are needed, such as amino acids, fats and iron. Vitamin B_{12}, a member of the B group of vitamins (p. 63), which is found in meat and liver, is also essential for blood formation. It is absorbed through the wall of the small intestine with the assistance of a carrier protein – intrinsic factor – which is secreted by the wall of the stomach (Fig. 8.2). If the diet is deficient in vitamin B_{12}, if intrinsic factor is missing or if the small intestine is diseased, a severe anaemia may result. The replenishment of some white blood corpuscles and the platelets also takes place in the bone marrow.

White blood corpuscles

The white corpuscles found in the blood are of several kinds, with different microscopical features and functions. They are mainly concerned with the defence of the body against bacterial infection and the entry of foreign material. They do this by forming antibodies (p. 90) and by direct attack. Many of them are able to engulf and destroy foreign particles, whether bacteria or debris from damaged cells and tissues. This property is called *phagocytosis*.

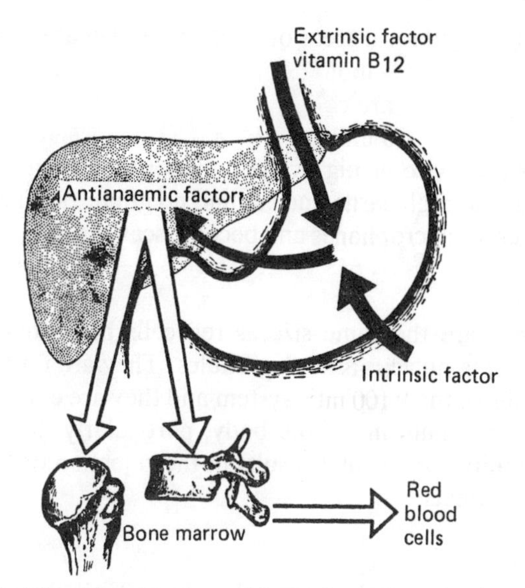

Fig. 8.2 The factors concerned in the formation of red corpuscles

White blood corpuscles collect in very large numbers around any area of local infection or tissue damage. Pus consists of masses of dead white blood corpuscles and liquefied, dead tissue.

Only three of the several different kinds of white blood cell will be dealt with here, the *polymorphonuclear leucocyte*, the *monocyte* and the *lymphocyte*. There are normally 5000–10 000 white cells per mm^3 of blood but the number varies during the day. They live for seven to fourteen days and there is a rapid turnover with constant destruction and replacement. There is usually a rise (leucocytosis) after meals, strenuous exercise or other forms of stress, such as infection. 70–75 per cent of the leucocytes are polymorphonuclear, less than 5 per cent are monocytes and the rest are lymphocytes.

Polymorphonuclear leucocytes

These cells are nearly twice as big as the red cells. They have granules in their cytoplasm and a nucleus divided into several lobes. They are formed from primitive cells in the lymph glands, spleen and bone marrow and are concerned with the reaction of the body to infection by bacteria and to the presence of foreign material in the tissues. They can engulf bacteria and other foreign particles and they collect around local areas of infection or damage in very large numbers.

Monocytes

Monocytes are twice as big as the polymorphonuclear leucocytes but they have a single large nucleus and no granules. They migrate to the tissues as macrophages and become scavengers.

Lymphocytes

Lymphocytes are the same size as red cells but contain a large densely staining nucleus and no granules. They are formed by the reticular cells of the lymphatic system and they are concerned with the defence mechanisms of the body, particularly the cellular or tissue immunity (p. 29) that results in the rejection of transplants from other animals.

Platelets

Platelets are tiny, granular corpuscles about 2 μ in diameter which are formed from special bone-marrow cells called *megakaryocytes* and are concerned in the control of bleeding after injury. There are usually 250 000–500 000 in every mm^3 of blood and their presence is essential for the normal retraction of blood clot which expresses the serum (p. 85). The platelets break up in areas of damage to the blood vessels and liberate serotonin, which causes small blood vessels to contract strongly, so tending to reduce bleeding after injury.

Control of bleeding

Loss of blood after injury to a blood vessel is reduced or stopped in three ways:

1 Contraction of the vessel wall;
2 Plugging of the hole by platelets;
3 Formation of blood clot.

The clotting of blood results from a series of chemical reactions involving numerous enzymes which is started off by thromboplastin released by damage to tissues or the breakdown of platelets. Eventually the fibrinogen of the blood is converted into threads of fibrin which form a solid mass, entrapping red and white blood corpuscles. In time this blood clot contracts and yellow serum – blood plasma less fibrinogen – is squeezed out (Fig. 8.1).

Blood transfusion and blood groups

If blood from one person is injected into the circulation of another, ill effects or even death may result. The defence mechanisms of the receiver cause the donated red corpuscles to form solid masses, which block the blood vessels and cause severe tissue damage, especially to the kidneys. The essential feature of this kind of defence mechanism is that a reaction occurs when a particular foreign protein (*antigen*) meets the *antibody* that is specific to it. In the case of red blood corpuscles the antigen-antibody reaction makes them stick tenaciously together.

There are two important antigens in human red blood cells, called A and B. An individual may have pure A corpuscles, pure B, both A and B, or O (no antigens at all). In addition, an individual has antibodies to the antigens not present in his own red blood corpuscles. The presence of matching antigen and antibody in the same person is not compatible with life because clumping would occur straight away. All human beings can be grouped by means of a simple test into one of four groups, related to certain mucopolysaccharides on the surface of the red blood corpuscles. These groups are A, B, AB and O. Each group also possesses the antibody that does not react with its own protein (Table 8.2). Group A subjects have B antibody, Group B have A antibody, Group AB have no antibodies and Group O have both. Group A blood can be given to Group A recipients and, in theory, to Group AB, Group B blood to Group B recipients and so on. Theoretically, Group O blood can be given to anyone – they are Universal Donors – and Group AB can receive any blood because they have no antibodies. In practice,

Blood group	*Antigen*	*Antibody*
O	O	Anti-A; anti-B
A	A	Anti-B
B	B	Anti-A
AB	A and B	None

Table 8.2

careful direct testing is essential for proper safety in blood transfusion.

Rhesus factor

Another blood group, first discovered in experiments with Rhesus monkeys, has become important in recent years. Eighty-five per cent of white people, called Rhesus positive, belong to this blood group; the other 15 per cent do not. Rhesus-positive subjects can safely receive Rhesus-positive blood from other people, because they do not harbour any antibody to the Rhesus factor. If Rhesus-positive blood is given to a Rhesus-negative person, then antibodies to the Rhesus factor may be produced, rendering any subsequent transfusion of Rhesus-positive blood into that person extremely dangerous. In addition, a Rhesus-negative woman married to a Rhesus-positive man may conceive a Rhesus-positive baby. Antibodies to the Rhesus factor may then be produced in the mother and pass into the baby's circulation causing injury or death of the infant.

The reticulo-endothelial system

The cells of the reticulo-endothelial system line the blood and lymph sinuses of the bone marrow, lymphatic glands and spleen. Their functions are concerned with defence and blood formation.

The lymphatic system

The lymphatic system is the reserve drainage system for the tissue spaces. It deals with excessive tissue fluid and any particles too large to pass through the walls of the capillaries. Without lymphatic vessels the tissue spaces would gradually become choked with tiny particles of protein, fat globules, pigment granules, bacteria and bits of cell. This would quickly interfere with the free interplay of osmotic and hydrostatic forces (p. 34) that normally move fluid in and out of the capillaries. Lymphatic vessels in the wall of the small intestine have the additional function of absorbing and transporting the products of fat digestion from the intestine into the venous blood stream.

The lymphatic system consists of a branching network of thin-walled vessels ramifying throughout the tissue spaces. They are similar to blood capillaries, but their walls are so porous that large particles pass freely through. The tiny lymphatic channels converge and unite like the veins and possess similar one-way valves to prevent back flow. The pumping action of the muscles is similarly important in maintaining lymphatic flow. The lymphatic vessels finally end by opening into the venous system (Fig. 8.3). The largest lymphatic vessel – the thoracic duct – runs up through the chest and opens into the left subclavian vein. At intervals along their course the lymphatic vessels open into lymphatic nodes, which are important filters. Special cells lining the nodes engulf particles in the lymph and immobilise or destroy them. Bacteria may lodge in a node and grow, causing pain and swelling. Soot breathed into the lungs is trapped in the lymphatic nodes inside the chest. Lymphatic nodes are grouped in clumps at the knee, elbow, armpit, groin and neck, where they can be felt as the 'glands'. Larger collections are also found in the abdomen and chest.

Composition of lymph

Lymph is an overflow from the tissue spaces and it has almost the same composition as tissue fluid, but it is a little more concentrated. Both lymph and tissue fluid have essentially the same composition as the plasma of the blood. Tissue fluid, however, contains very little protein, but lymph contains 2–3 g per 100 ml compared with the 6–7 g per 100 ml of blood plasma. Lymph leaving the liver has an even higher concentration of protein (4.5 g per 100 ml). This results from the continual new formation of protein by the liver from amino acids absorbed from the intestine.

The spleen

The spleen is the largest mass of lymphoid tissue in the body. It is a solid organ, weighing 250 g (½ lb) and situated in the upper abdomen just beneath the diaphragm on the left side. It forms the main part of the reticulo-endothelial system in man. It is not essential to life and no serious ill-effects are found after splenectomy. The spleen consists of a vast network or sponge formed by a supporting framework of muscular strands and connective tissue fibres. The sponge is filled with masses of lymphatic cells and phagocytic cells of the reticulo-endothelial system. The arteries which enter it break down into smaller and smaller arterioles

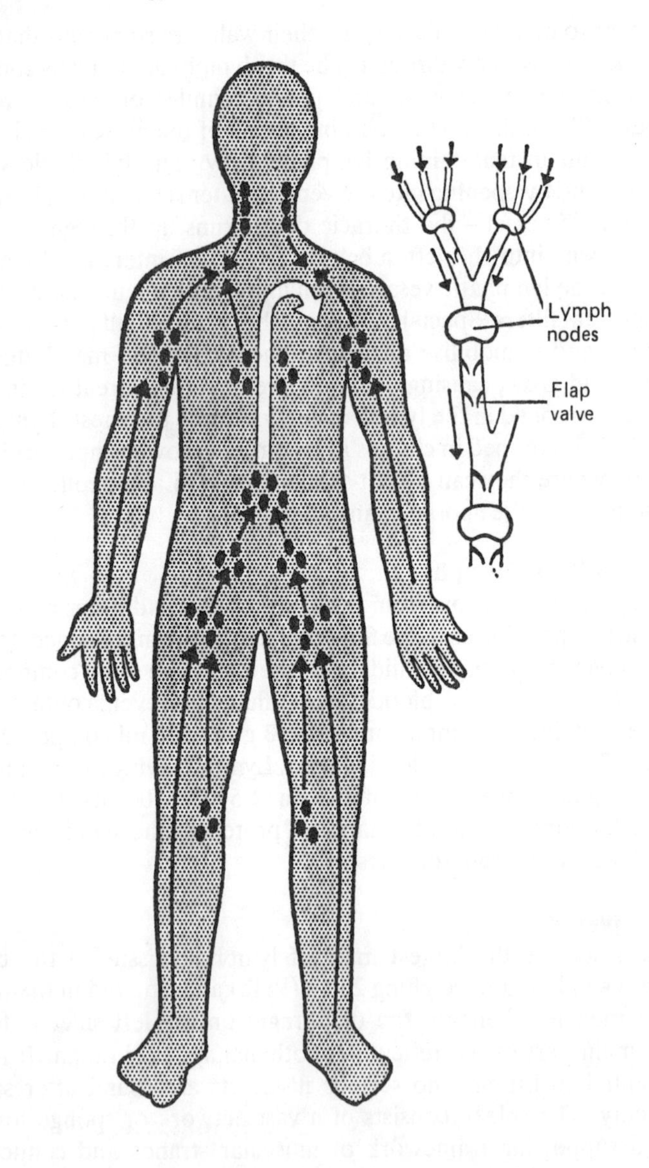

Fig. 8.3 The overall arrangement of the lymphatic system. The large white arrow represents the thoracic duct. *Upper right:* Detailed plan showing lymphatic ducts, lymph nodes and flap valves

opening into the mesh of the spleen itself, from which the blood is collected again through the venous sinuses.

FUNCTIONS OF THE SPLEEN

1 *Blood formation* The spleen is responsible for blood cell formation in the second half of foetal life and it redevelops this capacity in the adult if the bone marrow is destroyed. It also controls the maturation and escape of formed blood cell elements from the bone marrow in some way not understood.
2 *Blood destruction* The spleen removes old, damaged or deteriorated red cells and it has been described as the graveyard or slaughter-house of the blood.
3 *Reservoir* The spleen acts as a reservoir, particularly in animals, and when it contracts in response to exercise, asphyxia or haemorrhage, some of its contained blood is added to the circulation. This function is unimportant in man.
4 *Haemolysins* The spleen can form substances called haemolysins, which break down foreign red cells and sometimes the body's own red cells.
5 *Iron storage* The iron from haemoglobin of broken-down red cells is stored in the spleen and re-used later in fresh red cells.
6 *White cells* The Malpighian bodies form lymphocytes which are passed into the circulation.

Thymus

This soft pinkish-grey organ weighs 10–15 g at birth and is situated in the upper part of the chest in front of the trachea and aorta and deep to the sternum. At puberty it weighs 30–40 g but it then gradually atrophies. It is the chief organ of the lymphoid system. It controls the formation of lymphocytes, which form the essential basis of the cellular immune responses about to be described.

Immunity

The normal animal defends itself vigorously against any invasion by foreign materials. These are known as *antigens* and are usually proteins or polysaccharides. This built-in immunity enables the animal to resist most bacteria, parasites and fungi but also makes the transplantation of foreign organs, skin or tissues impossible without special techniques.

Each foreign organism or cell has a fixed, characteristic pattern of molecules on its surface (antigens) which the body can recognise and inactivate. Normally the body does not react to the antigens on its own cells. The agents responsible are lymphocytes.

Early in foetal life the precursors (stem cells) of all the lymphocytes found in the adult enter the foetus. Some enter the thymus gland and are called T-lymphocytes. The others go to the liver and spleen and become B-lymphocytes.

When a T-lymphocyte first meets a specific antigen it divides repeatedly forming large numbers of lymphocytes sensitised to react with the antigen on the foreign cell. It also produces T memory cells thus establishing immunity with an immediate response to any future invasion.

When the sensitised lymphocytes meet the invading cell they react to its surface antigens and forms a *lymphokine*, which breaks down its cell membrane and destroys it. This is the basis of cellular immunity against viruses, fungi, tumour cells, transplanted tissue and a few bacteria. In transplant surgery tissue is chosen with surface markers (i.e. antigens) closely resembling those of the recipient; close relatives or a twin are the best source. At the same time the T-lymphocytes of the recipient are blocked by an anti-lymphocytic serum, although this carries the penalty of an increased liability to infection.

The B-lymphocytes provide humoral immunity. They react to a specific antigen by producing large amounts of globulin antibodies which destroy foreign cells or neutralise their toxins (antitoxins).

Immunity may be natural or acquired. Natural immunity is inborn and depends upon the white blood cells and reticulo-endothelial system. Acquired immunity results from repeated invasion by foreign proteins, starting in infancy. More and more antibodies are produced with increasing age and susceptibility to infection tends to be lower in adults. This process can be reinforced by the deliberate administration of foreign proteins which are extracted by various methods from the organisms, e.g. diphtheria, poliomyelitis. Such procedures are called vaccination or inoculation and they produce *active immunity* which lasts for some years. *Passive immunity*, lasting for a few weeks, can be conferred by injecting the plasma of an immune person, containing the necessary antibodies, into the blood stream of a susceptible person.

It has been shown that the body may, occasionally, produce antibodies to its own tissue proteins when they escape from damaged cells. The antibodies then cause further cellular damage and a self-perpetuating cycle of damage may ensure. Certain diseases of the thyroid gland, liver and other organs may be so caused.

Allergy

Under certain circumstances harmful responses may occur when a foreign substance, or antigen, enters the body. These abnormal responses are called allergic and the individual is said to be hypersensitive. They can only occur if the person has already been exposed to the antigen at least once before and has developed some antibody to it. The type and severity of an allergic reaction depend upon the strength and persistence of the antibody screen evoked by previous exposure to the antigen. These antibodies are located on cells in the skin or mucous membranes of the respiratory and gastro-intestinal tracts. Typical antigens are pollens, dusts, feathers, wool, fur, some foods and drugs. The reactions causes the symptoms of hay fever, asthma, eczema, urticaria and contact dermatitis. If there is much cellular damage, excessive amounts of *histamine* may be released causing circulatory failure (anaphylaxis). In other cases antigen–antibody reactions of a different type inside the body may cause severe and lasting damage to blood vessels, tissues and organs.

9

The Kidneys

The kidneys form a functional unit with the ureters, bladder and urethra. Anatomically they are associated with the reproductive organs in the genito-urinary system. Functionally they are separate in the female and only distally linked in the male. The genital organs are therefore dealt with separately on page 187. The carbon dioxide, acids and other waste products of metabolism pass from the cells into the blood stream and are carried away to be discharged from the body. The carbon dioxide is eliminated through the lungs into the atmosphere and the other waste products are removed by the kidneys. Ideally, all the blood leaving the heart should pass through the kidneys before reaching the other organs. In practice, only a proportion of the blood pumped by each heart beat flows through the kidneys. However, every five minutes the amount of blood flowing through the kidneys is equal to the total volume of blood in the body. This arrangement does not ensure the complete removal of waste products but the amounts left are too small to injure even the sensitive cells of the nervous system.

There are two kidneys lying alongside the spinal column at the back of the abdomen at the level of the junction between the thoracic and lumbar vertebrae. They are 10 cm long, 6–7 cm thick and each weighs about 150 g. A short renal artery supplies each kidney with blood from the aorta under high pressure (Fig. 9.1) and the venous blood returns directly to the lower vena cava. The kidneys are buried in thick retroperitoneal fat and have tough fibrous capsules. They are firm and reddish-brown in appearance. When sectioned they show a thin rim or cortex and a medulla

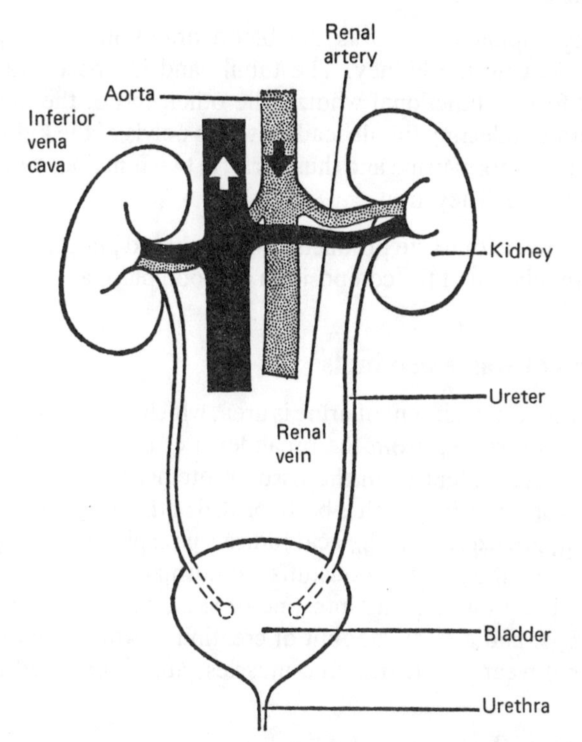

Fig. 9.1 The anatomical arrangement of the kidneys and urinary bladder

arranged in conical masses pointing inwards to the renal pelvis. From this a long tube – the ureter – runs down the posterior wall of the abdomen on each side into the pelvis. Here the ureters enter the base of the urinary bladder, where the urine is stored until it is discharged via the urethra.

Each kidney contains about one million functional units or *nephrons* consisting of elongated tubules, blind at one end and opening into a collecting duct at the other. At the blind end of the nephron there is a rounded expansion, called *Bowman's capsule*, which surrounds a cluster of capillary loops – the *glomerulus*. Each tuft of glomerular capillaries is supplied by an individual branch of the renal artery. Blood is led away from the glomeruli by special afferent arterioles which form further capillary networks surrounding different parts of the tubule of the same nephron. On leaving the secondary capillary networks the blood drains into the venous channels leaving the kidney. The tubule and its special vascular network form a functional whole. The other end of the nephron opens into a collecting tubule leading to the pelvis of the kidney.

The kidneys form urine and thus perform two functions which are essential to life. They are:

1 The removal of waste products from the blood plasma;
2 The regulation of the composition of blood plasma.

Removal of waste products

The chief solid constituent of urine is urea, which is formed from the amino acids resulting from the breakdown of dietary protein and also to a lesser extent from the tissue proteins. In addition small quantities of uric acid, which is both formed in the body and derived from food, are excreted. The acid radicals phosphate and sulphate from the metabolism of foodstuffs, abnormal substances which enter the body through the intestine or skin, such as alcohol and many drugs, and a small amount of creatinine, which comes from the normal wear and tear of the muscles, are all excreted in the urine.

Regulation of plasma composition

The kidneys play an essential part in the speedy and accurate maintenance of the osmotic pressure (p. 12), and electrolyte concentration, reaction and volume of blood plasma. If a large volume of water is drunk the kidneys rapidly excrete an equal volume of fluid with a low electrolyte content and a low specific gravity. In hot weather or in fever, where large volumes of fluid are being lost through sweating, the kidney produces a small amount of highly concentrated urine with a high specific gravity. The same accurate regulation applies also to the individual electrolytes – sodium, calcium, potassium etc. – which are conserved or excreted if their concentration in the plasma rises or falls above or below the normal values. Similarly the reaction of the plasma is influenced by the excretion of acid urine when there is a tendency to acidaemia (pp. 18, 98) and of an alkaline urine when alkalaemia develops.

Volume and composition of urine

The normal urinary volume is 1–1.5 litres per day, with a range of 0.5–3 litres per day. The amount formed depends upon the fluid and food intake, the climate and the amount of physical exertion. Less urine is formed in hot weather and after sweating, but excessive drinking of water, drugs and even the caffeine of coffee and tea may all increase the amount of urine secreted. Only half as much urine is formed during the night as during the day. The specific gravity of urine usually varies between 1.010 and 1.025, but the extreme limits are 1·002 to 1·040. The urine is normally acid (pH 6 approx.) but its reaction varies in relation to the needs of the body. Normal urine is pale yellow but it becomes darker if it is more concentrated as in fever. The yellow colour is due to pigments called urochromes, but traces of urobilin and other pigments are also present. When urine is allowed to stand it develops a smell of ammonia because the urea is broken down by bacterial action.

Normally 40–60 g of solids are excreted in the urine every day. Half of this (20–30 g) is urea and a quarter is sodium chloride (10–15 g). Urea is the chief end product of the breakdown of proteins and the amount present depends upon the diet and also upon the degree of tissue breakdown. The sodium chloride in the urine is proportional to the amount taken in the diet. In addition to urea and sodium chloride small quantities of creatinine, uric acid, sulphates, phosphates and amino acids are found. Proteins (albumin), glucose, bile and ketone bodies are found only under abnormal conditions.

Function of the nephron

The renal arteries carry blood at high pressure from the aorta into the glomerular capillary tuft. The blood pressure within the glomerular capillaries is 70–90 mm of mercury, which is between 50 and 70 per cent of the aortic pressure. The blood flow through the capillary tuft is controlled by the degree of contraction of the muscle of the arteriole leading to the tuft. The fluid pressure within the tuft forces some of the fluid part of the blood, by filtration, through the thin walls of the capillaries into the glomerulus and on into the tubule of the nephron (Fig. 9.2). The glomerular filtrate consists of blood plasma without the protein.

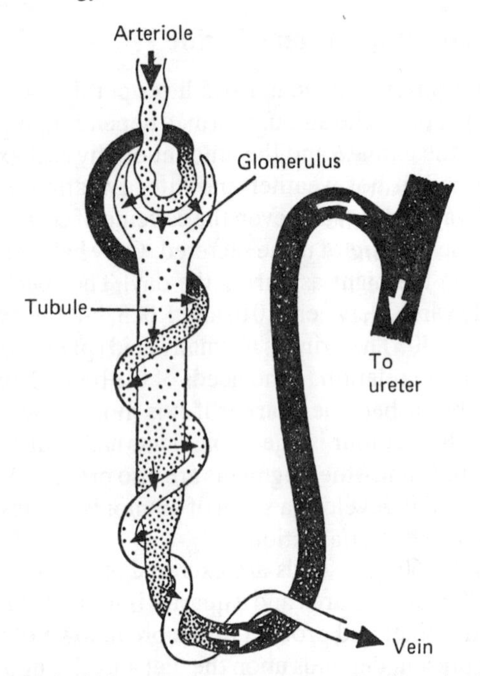

Fig. 9.2 A single kidney unit or nephron. Darker shading indicates greater concentration of excretory products

About 1200 ml of blood flow through the two kidneys every minute, which is almost a quarter of the total cardiac output. An amount equal to all the blood in the body, therefore, passes through the kidneys every four to five minutes. The renal blood vessels are not involved in the ordinary vasomotor reflexes (p. 161) and renal blood flow is maintained when the blood supply to other organs diminishes sharply or even stops in an attempt to maintain the circulating blood volume (p. 127). The pressure of 70–90 mm of mercury forcing fluid out through the glomerular capillary walls is opposed by the osmotic pressure of the plasma proteins (25 mm of mercury) and by the tissue pressure in the kidney itself (20 mm of mercury). If the effective glomerular pressure falls below this combined level of 45 mm of mercury then no urine is formed at all. This occurs when the systolic blood pressure falls below 80 mm of mercury, or if the pressure in the collecting system rises too high because of outlet blockage as in prostatic hypertrophy.

The total amount of glomerular filtrate is about 180 litres per day, yet the amount of urine formed from it is only 1–1.5 litres. Very large amount of water, and other substances, are reabsorbed by the kidney tubules. This is partly an automatic process because the blood leaving the glomerular capillaries is highly concentrated and exerts a strong osmotic pressure in the capillary network around the tubules drawing into the capillaries water and Na^+. About 80 per cent of the water and Na^+ in the filtrate moves into the renal medulla. The tubule now turns back on itself through the renal medulla and enters the collecting tubule. The filtrate is still isotonic with blood plasma in regard to sodium and water but contains a higher concentration of waste products. At this point 95 per cent of the water is absorbed from the tubule into the medulla, which is hypertonic. The amount removed is precisely controlled in relation to the water requirement of the body by antidiuretic hormone (p. 99) from the posterior pituitary gland. As a result concentrated or dilute urine is formed.

The absorption of electrolytes such as sodium and potassium is partly controlled by aldosterone and cortisol secreted by the supra-renal gland (p. 184) and the concentration of others, like chloride and bicarbonate, is related to the acid-base balance (p. 16).

Some of the reabsorption from the glomerular filtrate is a passive, automatic process of diffusion (p. 11) depending upon pressure gradients. This applies to water itself and in part to its electrolytes such as sodium, chloride, bicarbonate and calcium. These substances are all essential for the normal functioning of the body and they are absorbed in amounts sufficient to maintain their proper concentrations in the blood plasma. Other substances such as urea, phosphates and sulphates are waste products of metabolism unwanted by the body. They do not readily diffuse back into the plasma and a large proportion is excreted in the urine. The tubules are selectively porous to substances of value to the body and impermeable to the unwanted substances. As the filtrate passes down the tubules the concentration of waste products rises steadily and the specific gravity of normal urine varies from 1.015 to 1.030 compared with 1.010 for the glomerular filtrate.

Many substances in the glomerular filtrate show little tendency to diffuse through the tubular walls but still need to be conserved by the body. Such substances, e.g. glucose and amino acids, are returned to the plasma by a process of *active reabsorption*, which also applies to the electrolytes at times. Normally all the glucose in the glomerular filtrate is taken up and healthy urine contains no sugar. If too much sugar is presented to the renal tubules, i.e. when the blood sugar level rises to more than 10 mmol/l (180 mg per cent), reabsorption becomes inadequate to clear the glomerular filtrate and sugar will appear in the urine. Glucose is said to be a 'threshold' substance because it appears in the urine only when the blood sugar level rises above the threshold of 10 mmol/l (180 mg per cent). Sodium, potassium, calcium and chloride have low thresholds and are reabsorbed by the tubules to the degree necessary to mantain the correct osmolarity and electrolyte content of the body fluids. Waste products such as urea, uric acid and sulphate are also low-threshold substances, which are only reabsorbed by the tubules to a small degree. The excretion of phosphate is influenced by the hormones *parathormone* and *calcitonin*.

Regulation of acid-base balance

The metabolic activities of the cells cause a constant pouring of acids into the tissue fluids, where they are first accommodated by the buffering mechanisms (p. 17). The most important acids are lactic and phosphoric although sulphuric, uric and keto-acids are present in small quantities. Lactic acid, in the form of lactate, is dealt with by the liver but the other acid radicals produce a tendency to acidaemia which is counteracted by the excretion of acid urine by the kidneys. Sometimes the extracellular fluid becomes too alkaline and the kidney then removes base from the body by excreting alkaline urine. The pH of the urine varies widely in these processes. When the tissue fluids are acidaemic the pH of the urine may fall to 5.0, and in alkalaemia it may rise to 8.0, but the pH of the tissue fluids remains constant between narrow limits.

Water balance

The total blood flow through the kidneys is about 1200 ml per minute and the total extracellular fluid amounts to about 15 litres. The blood plasma and the extracellular fluid are in equilibrium with each other and therefore an amount of blood equivalent to almost all the extracellular fluid can pass through the kidneys every 15 minutes. The water and electrolyte content of the blood plasma and, indirectly, of the extracellular fluid are closely controlled by the kidney. Any increase of the water content of the body, such as a water load by drinking, leads to the prompt production of a corresponding amount of urine with a low specific gravity. This is brought about by slightly reducing the 99 per cent reabsorption of glomerular filtrate which normally occurs in the renal tubules.

Water absorption in the tubules is controlled by the antidiuretic hormone (ADH) of the pituitary gland (p. 181), which modifies the amount of water taken up in the distal part of the renal tubule. The secretion of this hormone depends upon the osmotic pressure of the blood entering the brain through the carotid arteries. An increase in the amount of water in the body lowers the osmotic pressure of the carotid blood and reduces the production of ADH, which causes an increased secretion of urine. The electrolyte content of urine is also partly under the control of a suprarenal gland hormone (aldosterone), which increases the uptake of sodium and indirectly of water in the renal tubules. Normally 99 per cent of the sodium in the glomerular filtrate is reabsorbed, but when the suprarenal glands are damaged or destroyed (Addison's disease) the amount of sodium in the urine increases and the amount in the blood plasma falls.

The kidneys and blood pressure

Persistently raised blood pressure is found in some patients with chronic renal disease. The mechanism is not clear but it is known that the kidney produces an enzyme called renin when its blood flow falls. Renin converts a globulin in blood plasma into angiotensin, which causes intense vasoconstriction with a resulting rise in blood pressure. It also releases aldosterone from the suprarenal cortex. This leads to retention of sodium and water so that the renal blood flow improves. Derangement of these mechanisms could be concerned in the development of high blood pressure.

Micturition

The urine which collects in the pelvis of the kidney on each side enters the corresponding ureter running down the back wall of the abdomen. The two ureters empty the urine into the muscular walled urinary bladder, which empties periodically through an outlet tube, the urethra. This movement is due to waves of muscular contraction (peristalsis) which travel down the ureter from kidney to bladder about three times per minute. The entrance to the urethra is normally sealed off by a ring of muscle – the *urethral sphincter*. When the bladder is full signals are passed down the motor nerve fibres from the spinal cord, making its muscular walls contract and relaxing the urethral sphincter. In young children urination is an automatic, reflex process, but it soon comes under conscious control. In the male the urethra passes through a large gland, the prostate, before running inside the penis (Fig. 16.1). In the female the urethra is short and runs straight down from the bladder, in front of the vagina, and opens into the perineum, the region at the top of the thighs (Fig. 16.2).

When urine begins to accumulate in the bladder its wall relaxes to accommodate the fluid so that the pressure inside remains at about 7–10 cm of water. When the bladder contains more than 400 ml of fluid its muscle begins to contract. This is a reflex response to the tension which develops in the bladder wall. The sensory pathway is uncertain but the reflex centre is in the sacral segments of the spinal cord and the motor pathways is the parasympathetic nerve supply of the bladder. When the amount of fluid is over 600 ml or the pressure is over 15 ml of water the contractions increase and pain may result.

Micturition is still under voluntary control because the reflex response to the rising pressure can be overridden or inhibited by the individual. Eventually the bladder wall contracts vigorously and the muscles of the perineum relax, while the muscles of the diaphragm and abdominal wall contract. The pressure in the bladder rises and the bladder neck slowly opens. Finally the external sphincter relaxes and urine is voided. Contraction of the bladder wall is normally continued until the bladder is empty; it then relaxes and the external sphincter and bladder neck close.

10

The Respiratory System

A continuous supply of oxygen must be delivered to every cell in the body. A short break in the oxygen supply disturbs the function of most cells and few can survive long without it. The cells of the nervous system, in particular, die in a few minutes if deprived of oxygen. The gas is taken up from the atmosphere through the lungs and is transported by the blood to all parts of the body. At the same time, the carbon dioxide constantly produced by the oxidation of foodstuffs is carried back to the lungs and discharged into the atmosphere. The gas exchanges in the lungs are called *external respiration* and those in the tissues are called *internal respiration*. The outside air contains one part of oxygen to four parts of nitrogen, with small amounts of carbon dioxide and traces of other, inert, gases. The carbon dioxide level in the outside air never rises too high because it is constantly used up by plants to produce glucose; oxygen is released at the same time.

Anatomy

The nose is divided by a partition of cartilage and also communicates with resonating cavities (sinuses) in the forehead and cheekbones. The air is filtered and moistened by ridges covered with mucous membrane which project into the nose. The back of the nose leads to the *pharynx*, from which open out the air and food pipes – the *larynx* and oesophagus. The larynx is closed off during swallowing by a flexible plate of cartilage – the *epiglottis* (Fig. 7.2). The larynx is a cartilaginous box which contains the vocal cords. It

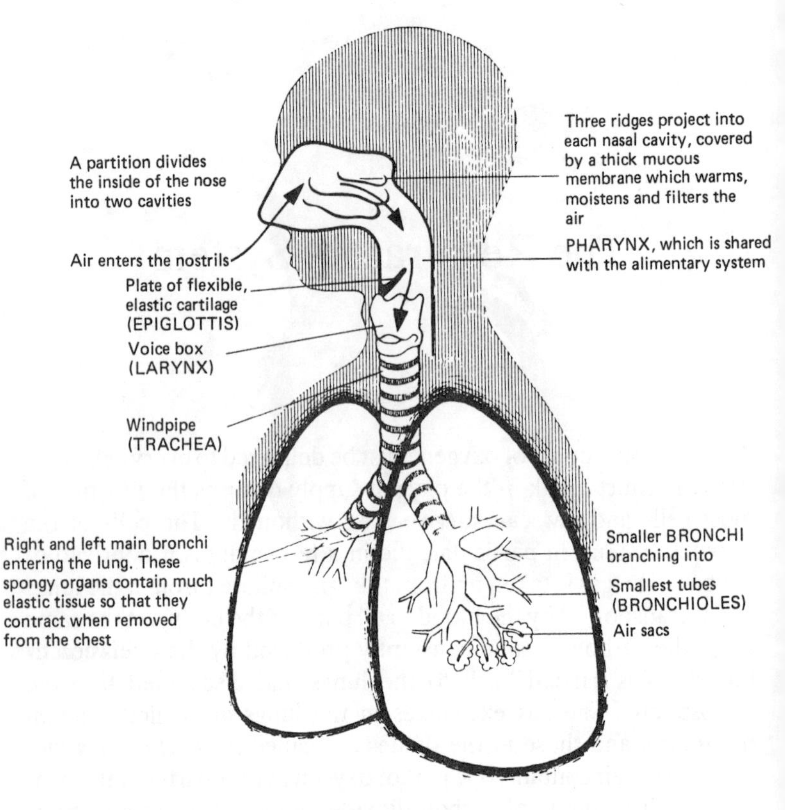

Fig. 10.1 The airways

leads into the windpipe (*trachea*), which divides into right and left main *bronchi* leading to the two *lungs* (Fig. 10.1). The trachea and larger bronchi are kept open by rings of cartilage set in their walls. The bronchi branch continuously into smaller and smaller subdivisions – *bronchioles* – and finally end in tiny air sacs – the *alveoli*. The lung structure can be likened to a bunch of grapes, where each grape is an alveolus, the stalks are the bronchi and bronchioles; the exchange of gases occurs through the skin of each grape. Each alveolus is extremely tiny, but because of the immense number present they comprise an area of about 90 m^2, which is the size of half a tennis court. The alveoli are thin-walled spaces surrounded by a network of tiny blood vessels – capillaries. The walls of the air

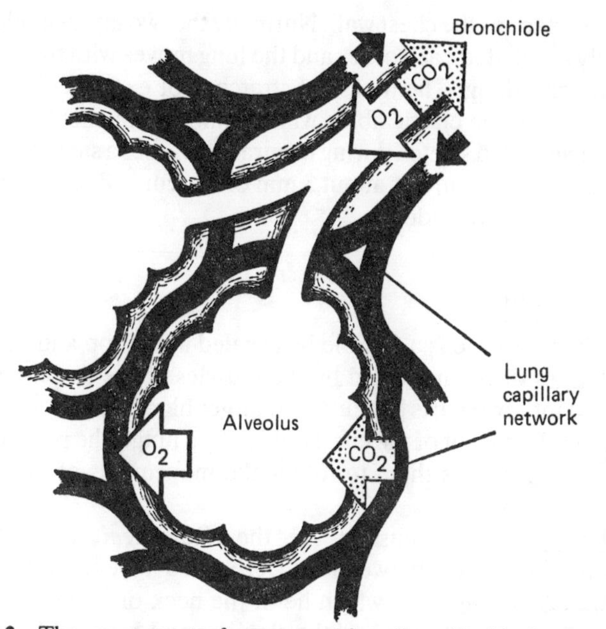

Fig. 10.2 The movement of oxygen and carbon dioxide between the capillaries and the atmosphere in the alveoli of the lungs

sacs and the capillaries are so thin that oxygen and carbon dioxide can easily pass through (Fig. 10.2).

Blood leaves the right ventricle of the heart by the pulmonary artery and is distributed to the lungs through a system of branching tubes which eventually leads into the pulmonary capillaries surrounding the alveoli. At this point only two thin layers of cells separate the air from the blood and gases can diffuse freely between them. As it traverses the alveolar capillaries the haemoglobin of the blood takes up oxygen to its full capacity (97 per cent saturated) and carbon dioxide passes out of the blood into the alveolar spaces. The alveolar air tends towards gaseous equilibrium (having the same tension of gases) with the blood which is constantly passing through the lungs. The alveolar air must be regularly changed by breathing if the supply of oxygen and removal of CO_2 are to be properly maintained.

The lungs are covered by a smooth glistening membrane called the *pleura*, which turns back at the root of each lung and covers the inner surface of the chest wall. Normally the two layers of pleura are closely applied to each other and the lung moves with the chest wall during breathing. The lungs are stretched or expanded against the resistance of the elastic fibres which ramify throughout their structure. The added stretch during inspiration causes a small decrease of intrapleural pressure of about 3 mm of mercury which pulls air into the lungs from outside.

Ventilation

The thoracic cage is a flexible box sealed at the top and bottom by the tissues of the neck and by the muscles of the diaphragm. It is divided into two parts by the heart, larger blood vessels and roots of the lung. Each half of the chest is filled by a lung. The pressure of air in the lungs forces them to follow the movements of the ribs and diaphragm exactly.

The respiratory muscles are the *diaphragm*, the *intercostal muscles*, which lie between and are fixed to each rib, and the *sternomastoid muscles*, which lie in the neck on each side but are attached to the sternum and the clavicles and help to lift the upper chest in strong respirations. The diaphragm is a thin sheet of muscle which arises from the body wall at the junction of the ribs and abdomen, and separates the thorax from the abdominal cavity.

On inspiration, the intercostal muscles contract and pull the ribs upwards and outwards on their hinges at the spinal column. At the same time, the diaphragm contracts and becomes less domed (Fig. 10.3). This enlarges the capacity of the chest, the lungs must expand and air is actively drawn into them through the trachea – inspiration. The lungs expand with the walls of the thorax, air flows into the trachea, bronchi and bronchioles, their walls stretching slightly, and the alveoli become filled with air. On expiration, the thoracic muscles relax, the thorax decreases in size, the elastic fibres of the lungs contract to their resting length and air is driven out of the chest. This process is repeated fifteen to twenty times per minute in the resting person, much more often during exercise. During quiet breathing the ordinary intake of air or *tidal volume* is about 500 ml (0.5 litre). The total amount of air that can be expelled after a deep inspiration is the *vital capacity* and amounts to 3.5–5.0 litres. Even then about 1.5 litres of air remains in the lungs forming the *residual*

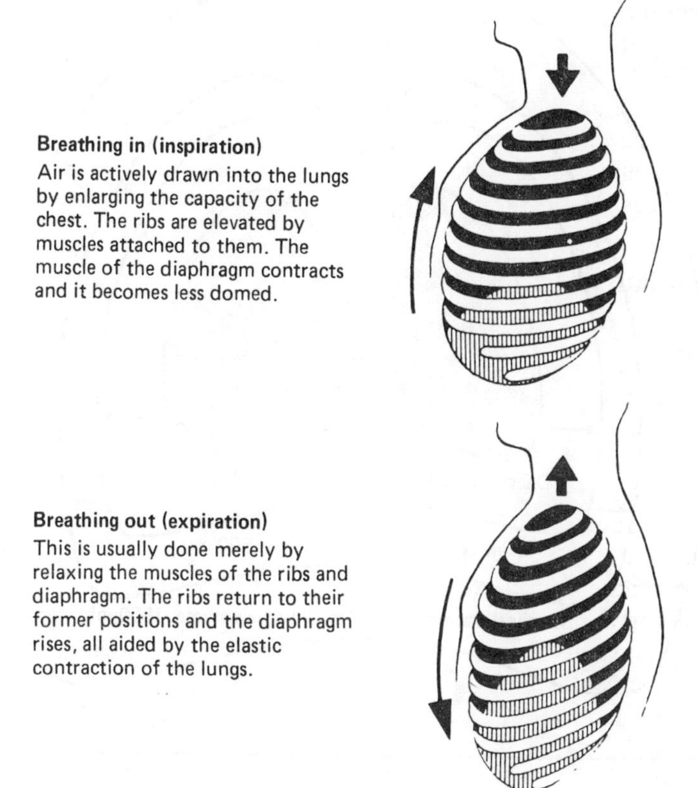

Fig. 10.3 The mechanics of ventilation of the lungs

volume (Fig. 10.4). Atmospheric air contains oxygen (O_2) 20.9 per cent, nitrogen (N_2) 79 per cent and carbon dioxide (CO_2) 0.03 per cent. The composition of alveolar air is O_2 14 per cent, N_2 80.5 per cent, CO_2 5.5 per cent, but it is saturated with water vapour.

Transport of gases

In each minute, under normal conditions, about 250 ml of oxygen are taken up and 250 ml of CO_2 are given out by the body, and these are the amounts of the two gases which enter and leave the blood in the lungs. Similar exchanges occur in reverse in the tissues, where oxygen is given up and CO_2 is removed. The exact amount of CO_2

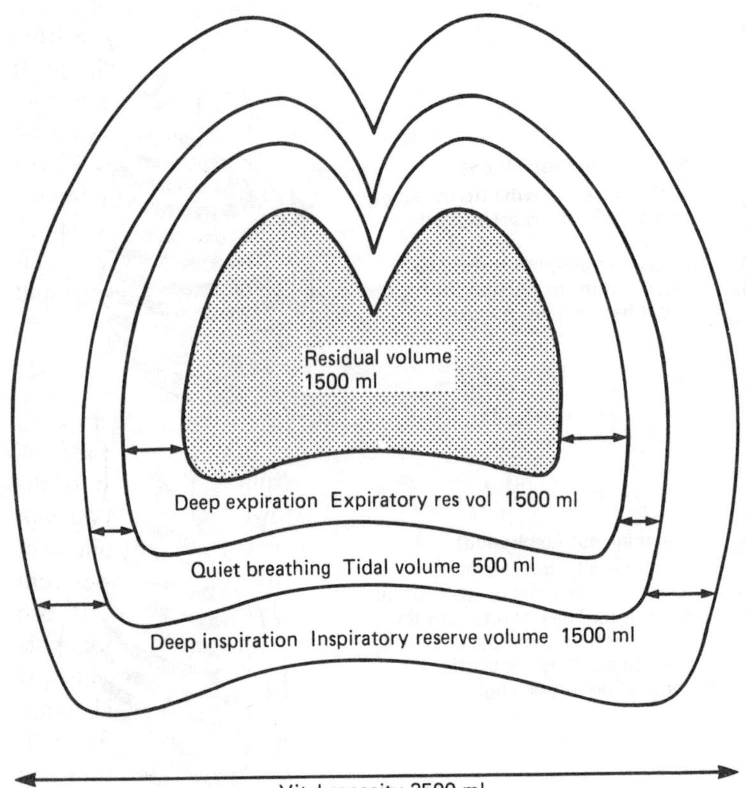

Fig. 10.4 The subdivisions of the lung air

expired depends upon the metabolism (p. 10), the acid-base balance (p. 16) and the pattern of respiration. Blood from the right side of the heart contains less oxygen and much carbon dioxide. It passes along the pulmonary arteries to the capillary network around the air sacs. Gases must pass from a region of high concentration to a region of low concentration by diffusion. In the air sacs oxygen passes readily from the air into the oxygen-poor blood returning from the tissues. The oxygen combines with the haemoglobin in the red blood corpuscles and is carried to the tissues. The blood reaching the lungs contains much carbon dioxide, which in the same way passes out of the blood into the air sacs. In the tissues the reverse process occurs, with each gas diffusing from regions of high to low concentration. The overall effect is to deliver oxygen from the air to the tissues and carbon dioxide from the tissues to the air. If the cells of the body use up more oxygen during activity, then more diffuses out of the blood into the tissues. The extra carbon dioxide produced by the activity enters the blood and is carried away. Nerve cells at the base of the brain respond to the rise in carbon dioxide concentration in the blood by sending out nerve signals which increase the rate and depth of respiration. In this way breathing is continually regulated so that the needs of the body for oxygen input and carbon dioxide output are always fully met.

Oxygen carriage by the blood

The liquid part of the blood (the plasma) is a very poor carrier of oxygen because, at the pressures available, only 0.3 ml of oxygen can dissolve in 100 ml of plasma, which is quite insufficient for the needs of the body. The red blood cells, however, contain *haemoglobin* (p. 82), a protein which can combine with a large volume of oxygen so quickly that in the lungs it may become 97 per cent saturated, forming a compound called *oxyhaemoglobin*. It also forms specific compounds with other gases such as carbon monoxide and carbon dioxide. One gram of haemoglobin combines with 1.34 cm^3 of oxygen and 100 cm^3 of normal blood contains 15 g of haemoglobin, which can combine with 20 cm^3 of oxygen ($15 \times 1.34\ cm^3$). This is called the oxygen-carrying capacity of the blood. The combination of haemoglobin and oxygen is also affected by the acidity (pH) and by the carbon dioxide content of the blood. Any increase in the carbon dioxide content of blood tends to reduce the amount of oxygen taken up by haemoglobin at any given partial pressure of oxygen. Similarly an increase in acidity in the blood (fall in pH) tends to drive oxygen out of combination with haemoglobin. In the lungs carbon dioxide diffuses out of the blood into the alveolar air, with the result that there is an increased tendency for oxygen to be taken up by the haemoglobin. In active tissues there is a low oxygen tension which allows the diffusion of oxygen from blood to tissue. Similarly the high partial pressure of carbon dioxide in the tissues makes carbon dioxide pass into the blood, where it assists in driving oxygen out of combination with haemoglobin. In these ways oxygen is taken up in the lungs and discharged in the tissues.

Carbon dioxide carriage

The passage of carbon dioxide into and out of the blood in the tissues or lungs depends upon the same mechanisms of gradients of partial pressure and diffusion as does the movement of oxygen. Every 100 cm^3 of whole blood contains approximately 50 cm^3 of carbon dioxide. Five per cent of this is in solution and 95 per cent is in various forms of loose combination with chemical compounds.

The carbon dioxide formed by tissue metabolism diffuses into the blood plasma and from there passes into the red blood cells. In the red cells it reacts with water under the influence of the exzyme carbonic anhydrase to form carbonic acid.

$$CO_2 + H_2O \rightleftharpoons H_2CO_3$$

The carbonic acid immediately ionises into H^+ and HCO_3^- ions. Most of the hydrogen ions combine with haemoglobin in the red cells and with plasma proteins to form un-ionised compounds so that they do not increase the acidity of the blood (buffering). The bicarbonate ions combine with the base in the cells (K^+) and to a lesser extent in the plasma (Na^+) and are transported as bicarbonate to the lungs.

About one-fifth of the carbon dioxide which enters the red cells combines with reduced haemoglobin to form carbaminohaemoglobin. This compound tends to break down again in the presence of oxygen so that when oxyhaemoglobin is formed in the lungs there is the reverse movement of carbon dioxide away from haemoglobin. When the venous blood reaches the lungs, the carbon dioxide diffuses out into the alveolar air and the chemical processes in the red cells are reversed. Under the influence of carbonic anhydrase in the red cells the bicarbonate ions (HCO_3^-) tend to combine with the hydrogen ions (H^+) to form carbon dioxide and water. This is the reverse of the process which occurred in the tissues. The carbon dioxide diffuses out into the alveolar air and more bicarbonate moves into the red cells from the plasma while chloride moves back from the red cells to the plasma.

In this way there is a delicate balance between the absorption and excretion of oxygen and carbon dioxide in the lungs and tissues respectively. The changes which occur can be summarised in the following way:

In the tissues

1 Oxygen leaves the red cells.
2 CO_2 enters the red cells.
3 Bicarbonate ions leave the red cells and enter the plasma.

In the lungs

1 Oxygen enters the red cells.
2 CO_2 leaves the red cells.
3 Bicarbonate ions enter the red cells from the plasma.

The control of respiration

The very delicate and sensitive control of the tension of carbon dioxide in the arterial blood is the responsibility of the respiratory centre in the medulla of the brain. This is not a centre in the anatomical sense but is a diffuse collection of nerve cells which coordinate respiratory activity. They seem to have an intrinsic rhythmic property but they are also affected by impulses travelling in the vagus and glossopharyngeal nerves and by changes in the composition of the blood circulating through the medulla. The respiratory centre influences respiration through connections with nerve cells, in the third and fourth spinal cord segments, giving rise to the phrenic nerves to the diaphragm and with the anterior horn cells supplying motor nerve fibres to the intercostal and other respiratory muscles.

Chemical control

The chief chemical factors in the control of respiration are the carbon dioxide tension and the H^+ ion concentration of the blood circulating through the medulla. Any increase of the carbon dioxide tension leads to an increased depth of respiration so that more carbon dioxide is excreted in the expired air. Similarly an increase in the blood H^+ ion concentration (fall in pH) causes increased rate and depth of respiration with consequent elimination of carbon dioxide and an eventual fall in the blood pH. Decrease in carbon dioxide tension or H^+ ion concentration causes a corresponding reduction in respiratory activity. Lack of oxygen has little effect upon the respiratory centre until the blood oxygen tension falls well below 40 mm of mercury. The length of voluntary breath holding which is possible depends on the rate of rise in the carbon dioxide tension of the blood; above a certain point respiration occurs involuntarily. The increased respiratory rate in exercise is largely due to the increased production of carbon dioxide in the tissues, which eventually stimulates the respiratory centre.

Nervous control

The chemical control of respiration is supplemented by nervous control from several sources. Nerve endings in the carotid and aortic bodies are sensitive to oxygen lack and send signals to the respiratory centre through the nerve fibres in the glossopharyngeal nerve. There are also receptors in the small bronchioles which are sensitive to distension of the lung during inspiration. They send impulses by way of the vagus nerve to the respiratory centre, which cuts off inspiration when the degree of stretch becomes too high. In this way the rhythm of respiration is controlled by nervous influences, while its depth tends to be controlled by the carbon dioxide concentration of the blood.

Effects of exercise

The responses depend on the type of work and the fitness of the individual. The trained athlete has a greater working capacity and endurance. With exertion there is a gradual increase in oxygen consumption which stabilises after a few minutes. Pulmonary ventilation rises from 5 litres per minute at rest to 100–150 litres per minute. The respiratory rate rises, often in time with rhythmical exercise as in running, almost entirely by shortening the expiratory phase while the tidal volume increases 3–4 times. There are profound changes in the gas content of the venous blood, with a fall in the oxygen level and a rise in carbon dioxide. However, gas exchange in the healthy lung is so efficient that the gas content of the arterial blood leaving the lungs remains virtually normal.

Several factors contribute to these responses. Nerve impulses from muscles and joints plus conscious anticipation initiate them. Rising CO_2 levels stimulate the respiratory centre and its cells are sensitised to CO_2 by the consequent rise in body temperature and by oxygen shortage in very vigorous exercise. The rise in the level of lactic acid in the blood also acts as a direct stimulus. After exercise the respiratory responses slowly return to normal as any 'oxygen debt' (p. 48) is repaid and the respiratory centre regains its usual sensitivity.

Environmental influences

The discussion so far has presumed that the healthy subject is breathing normal air at sea level. In fact the air may be polluted or attenuated by altitude and the respiratory system may be damaged by disease.

As the atmospheric pressure falls on mountains or in climbing aircraft the oxygen pressure and availability fall to a danger point at a height of 5000 metres or more. If no extra oxygen is supplied consciousness is lost and permanent loss of brain cells, or death, may follow after a few minutes. In mountain climbing the changes are slower and some degree of acclimatisation occurs. This is largely through an increase in the number of circulating red blood corpuscles enabling more oxygen to be assimilated. Even so exertion without extra oxygen becomes increasingly slow, laborious and distressing above 6–7000 metres. If the oxygen supply is slowly reduced as in a sunken submarine there is a gradual decrease in physical capacity accompanied by increasing respiratory distress caused by the steady rise in CO_2 content. Eventually consciousness is lost and death ensues.

Workers in compressed air and deep-sea divers face additional problems. CO_2 becomes increasingly toxic at raised pressures and has to be removed. Nitrogen under pressure can cause a form of narcosis with irrational behaviour. In addition nitrogen is forced into solution in the body fluids. If the pressure is suddenly reduced the nitrogen forms bubbles which can block arteries and stop the flow of blood. This produces painful complications – the bends – and may cause permanent paralysis. After prolonged dives the return to normal pressures has to be very gradual and in some special cases divers may take a week or more to recover.

Air pollution

Industrial and urban development has caused increasing pollution of the air with a growing danger to health. This has been realised in recent decades and public health control measures have produced considerable improvement. Fog is a common natural event but if it is contaminated with dust and sulphur dioxide from the burning of fossil fuels by industry and in domestic fires it can be a serious danger to health by exacerbating existing chronic lung disease.

About 25 million tonnes of sulphur are produced annually in Europe from power stations, furnaces and refineries. This forms sulphuric acid which falls as acid rain. It reduces crop yields, damages trees, corrodes buildings and contaminates drinking water when it enters reservoirs and aquifers.

In some areas, such as Los Angeles in the USA, 'smog' has become a serious problem. It results from the photochemical oxidation of motor car exhaust gases in the air. Ozone and oxides of nitrogen are formed and affect lung function. The resulting oxygen lack can impair mental performance and sometimes produce permanent damage. The lead in petrol can also cause irreversible damage to the nervous system.

Although public health legislation has improved the quality of the air in many respects one important factor – cigarette smoking – is proving hard to control. There is now very strong evidence that smoking is the main cause of chronic bronchitis and lung cancer and it is strongly suspected to be an agent contributing to some cancers of the genito-urinary system and the intestine. In addition, the carbon monoxide in cigarette smoke is of major importance in the development of coronary artery disease and heart attacks. Even non-smokers may be at risk if they associate with cigarette smokers.

There are many examples of occupational air pollution causing disease among the workers or even in their families and among those living near the factories. The best known is silicosis due to inhaling silica, which is a major component of the earth's crust. Mining and allied industries are the chief cause of this crippling and often fatal disease. Iron, tin and coal miners develop closely allied lung diseases. Asbestos is a particularly dangerous pollutant. It not only produces chronic lung disease but twenty years after even a short exposure a very malignant form of cancer – mesothelioma – may develop in the pleura or peritoneum. Workers in many other industries from cotton mills (byssinosis) to chemical factories are at risk from a multitude of agents. Radioactive materials are a special danger to health because of the possibility of long-term malignant change. They are widely used in industry, research and medical practice but may also be accidentally released from nuclear reactors. Many authorities consider that all exposure to radioactivity is dangerous and deny the possibility of a safe level or threshold.

11

The Heart and Circulation

The cells of the body are ceaselessly active during life and their health depends upon the chemical and physical composition of the tissue fluid (p. 34) which bathes them and which maintains the correct conditions of temperature, reaction and nutrition. Tissue fluid is in equilibrium with circulating blood, which is the transport medium for the gases, nutrients, chemical messengers, defence materials and waste products. Blood is a sticky, red fluid that carries oxygen and food to the tissues, carbon dioxide to the lungs and waste products to the kidneys.

The heart and circulation perform this essential service of supply and removal in a manner which is delicately adjusted to meet the constantly changing needs of the body. The heart and blood vessels are a closed system with elastic walls; the blood never comes into direct contact with the cells. The tubes leading out of the heart are called *arteries* and have a smooth inner surface of flattened cells. The arteries continually divide into smaller and smaller branches. The smallest arteries (*arterioles*) branch into networks of very small vessels called *capillaries*. They have no muscle and their walls are so thin that gases and other chemicals travel freely through them. Large molecules, such as proteins, cannot pass. Capillaries run close to all the cells of the body. Except in the central nervous system (p. 133) all interchange of gases and chemical substances of all descriptions occurs through the walls of the capillary blood vessels into the tissue fluid that bathes the cells. There is a constant exchange of water and chemicals between the tissue fluid and the blood plasma. The tissue fluid is completely renewed every few minutes and in active organs, where the capillaries are dilated, the turnover of tissue fluid is much faster. The activity of the heart and circulation is delicately and precisely adjusted to the ever-varying needs of the different parts of the body. The blood moves constantly through the arteries to the tissues and back in the *veins* to the heart. The volume of blood supplied to a particular organ varies continually in relation to its level of activity and to the simultaneous demands of the rest of the body. At all times there is a strong tendency to maintain the circulation of, first, the brain and, secondly, the kidneys at the expense of the other parts of the body.

Cardiovascular anatomy

The heart consists of two similar, but separate, muscular pumps side by side – right and left – which can only communicate through two intricately branching systems of tubes (Fig. 11.1). The left-side pump leads into the *systemic vessels*, which carry oxygenated blood to all the tissues of the body and return it to the right-side pump. This gives rise to the *pulmonary vessels*, which carry oxygen-poor, carbon-dioxide-rich blood to the lungs. From them the refreshed blood returns to the left-side pump.

Although the two pumps lie side by side forming the heart, they are entirely independent units. Each pump has two chambers, an *atrium* and a *ventricle*. The muscle of the atria and ventricles is separated by a fibrous ring into which the muscle fibres are inserted. The *endocardium* is a smooth membrane lining the inside of these chambers. The *pericardium* is a membranous bag which forms an outer cover to the heart, permitting it to move freely within the chest as it pumps. Both chambers have walls of cardiac muscle, which has the special property of automatic, rhythmic contraction (p. 123). The atrium is a receiving chamber and has thin walls; the ventricle is the main pump and has thick walls. One-way, flap or pocket valves (Fig 11.2) are situated between the atria and the ventricles, at the outlets from the ventricles and in the systemic veins. They ensure that blood flows in one direction only. Any reverse flow fills out the flaps or pockets, pressing them together so that they become non-return valves, preventing the backflow of blood.

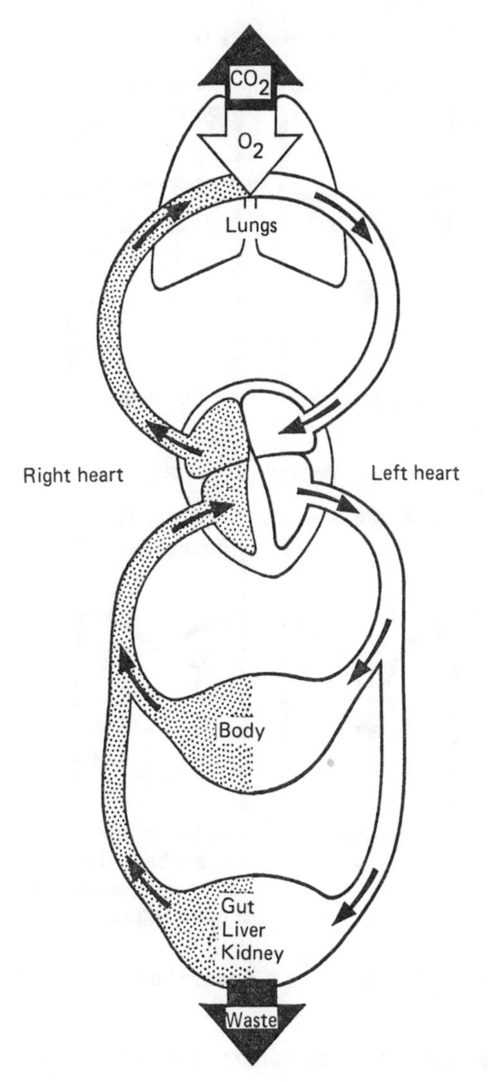

Fig. 11.1 Simplified diagram of the circulatory system. *Upper third:* Pulmonary circulation. *Lower two-thirds:* Systemic circulation

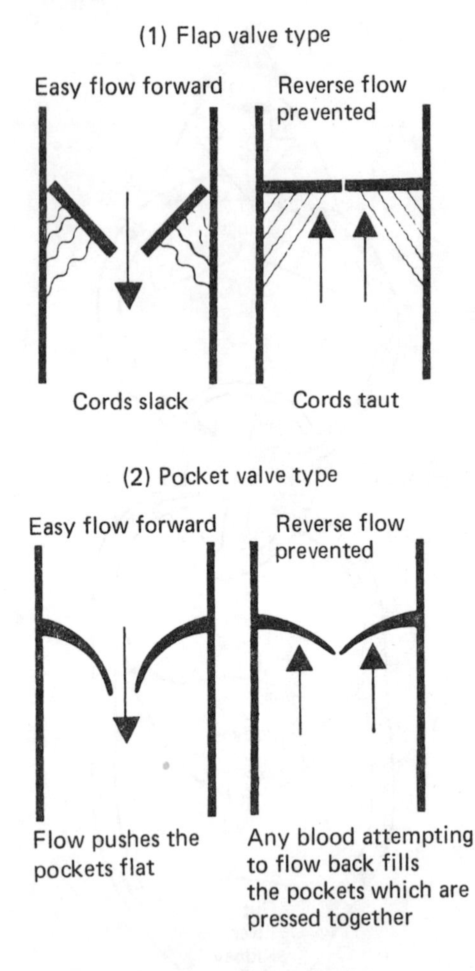

Fig. 11.2 Types of valves

There are no valves between the venous system and the atria, but each atrium is separated from its ventricle by a one-way valve called an *atrioventricular* (AV) *valve*. These valves consist of membranous flaps arising from the fibrous ring between the atrium and the ventricle. They ensure that the ventricles empty their blood into the pulmonary artery or aorta and do not squirt it back into the atria. The right atrioventricular valve has three flaps and is called the

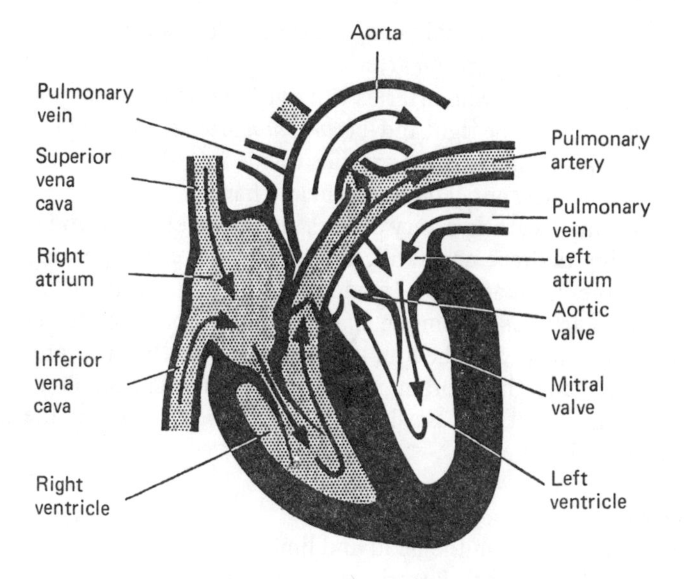

Fig. 11.3 Diagrammatic cross-section of the human heart

tricuspid valve; the left AV valve has two flaps and is called the *mitral valve* (Fig. 11.3). There are also one-way valves between the right ventricle and the pulmonary artery and between the left ventricle and the aorta. They are called the *pulmonary* and *aortic valves* respectively and they prevent the blood from running back into the ventricles during diastole.

Arteries

Each ventricle leads into a large artery directed upwards (Fig. 11.4). The pulmonary trunk from the right ventricle divides into the main right and left *pulmonary* arteries leading to the two lungs. The *aorta*, opening out of the left ventricle, gives off the main arteries to the arms and head – the *subclavian* and *carotid* arteries. It then turns to the left and arches downwards through the chest and diaphragm into the abdomen. In the chest the aorta gives off the *coronary* arteries, which supply the heart muscle itself and arteries which pass round the chest wall between the ribs on both sides. In the abdomen many branches spring from the aorta to supply the liver, stomach, intestines, kidneys and spleen. The aorta then divides into the two main *iliac* arteries, which run into the thighs as the right and left *femoral* arteries.

The two subclavian arteries run over the first rib and under the clavicle to form the right and left *brachial* arteries in the upper arms. At the elbow the brachial arteries divide into the *ulnar* and *radial* arteries, which supply the forearm and hand.

Each main carotid artery divides into an external carotid branch, supplying the face and skull, and an internal carotid branch, supplying the brain and eye. In addition, the brain is supplied by two *vertebral* arteries running in channels in the bones of the vertebral column in the neck. The four arteries supplying the brain are linked together beneath the brain by an arterial circle. The blood supply to the brain is not necessarily interrupted even if one of the main arteries is blocked.

Veins

Blood returns from the head and limbs in thin-walled veins, which usually run near to the arteries and have similar names. the veins are close to the surface at first but dive deeply to join the main veins as they approach the trunk. They have non-return valves along their course. The two iliac veins join to form the lower *vena cava*, which runs up through the abdomen and chest to enter the right atrium. In the abdomen it is joined by appropriately named veins that drain the blood from organs like the kidney. The blood from the intestine all runs into one large *portal* vein, which enters the liver. In this way the nutrients after absorption are taken straight to the main chemical factory and storehouse of the body. The blood from the liver travels in the *hepatic* vein to join the lower vena cava. Blood is drained from the head and neck in the *jugular* veins, which accompany the two main carotid arteries. They join with *subclavian* veins from the arms to form the upper vena cava, which enters the right atrium. The two main pulmonary veins from the lungs end in the left atrium.

Foetal circulation

The circulation of the blood in the foetus in the uterus is different from that in the adult because the placenta (p. 192) acts in place of the lungs and intestines. Blood, charged with oxygen and nutrients and freed from carbon dioxide and waste products, leaves the placenta by the umbilical vein (Fig. 11.5). This opens into the

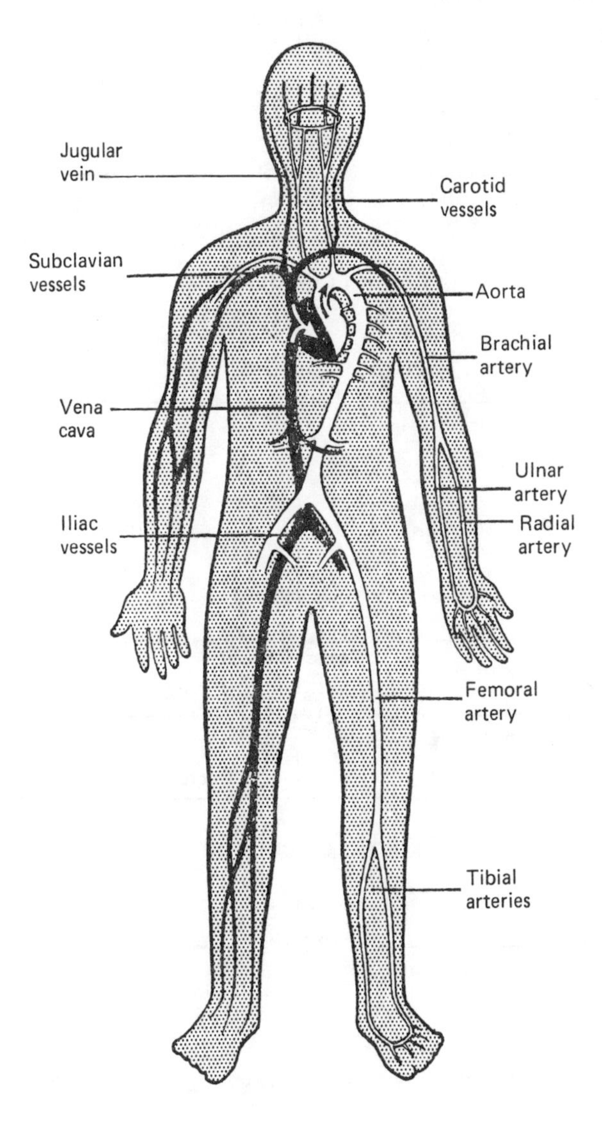

Fig. 11.4 The overall arrangement of the circulatory system. Veins – black. Arteries – white

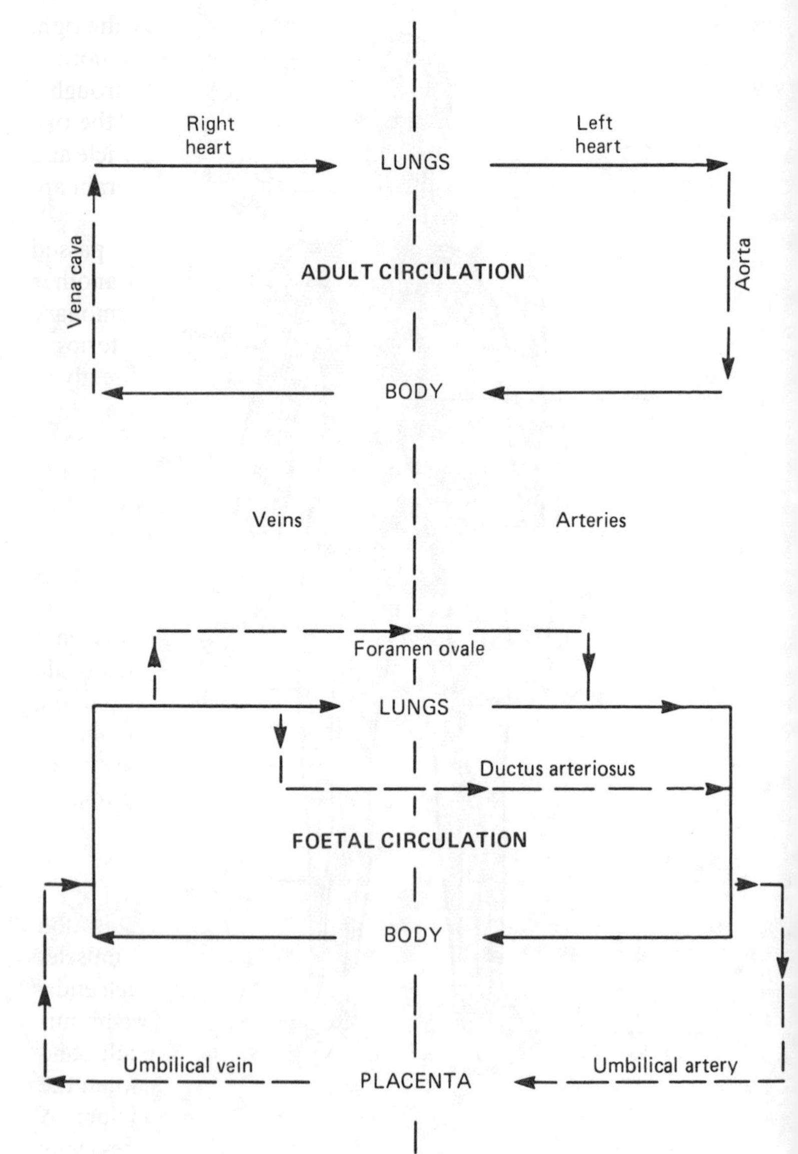

Fig. 11.5 Diagram of the circulation in the foetus and in the adult

inferior vena cava of the foetus and the blood passes on to the right atrium. Some of the blood enters the right ventricle in the normal way but part of it is shunted directly into the left atrium through a hole called the *foramen ovale* (Fig. 11.6), which connects the two atria at this stage of life. This blood then enters the left ventricle and is pumped out into the aorta, ensuring that the head and brain are well supplied with oxygenated blood.

A little of the blood which enters the right ventricle is passed through the lungs, but most of it by-passes the lungs in another shunt called the *ductus arteriosus*, which connects the pulmonary artery to the aorta. Blood passing through the ductus arteriosus misses the lungs and the left side of the heart but passes directly to the aorta. Some of the aortic blood passes through, and supplies, the body's organs and limbs, eventually returning to the right heart by the vena cava. The rest of the aortic blood passes along the umbilical arteries to the placenta, where it is refreshed and then returned to the foetus.

In the last month before birth increasing amounts of blood pass through the lungs and shortly after birth most of it does so. The placental vessels are cut after delivery thus excluding the placenta from the circulation. In the next few days both the foramen ovale and the ductus arteriosus are automatically sealed off and the adult pattern of circulation is established. Occasionally the ductus arteriosus fails to close rendering the circulation relatively inefficient unless the surgeon can close it at operation in later years.

Contraction of the heart

Cardiac muscle consists of bundles of cells with a basic organisation of actin and myosin molecules similar to that of skeletal muscle. They are about 80 μm long and 15 μm in diameter and at each end a specialised intercalated disc separates them from similar neighbouring cells. These discs serve as anchors for the actin molecules and also form junctions where electrical activity can flow between the cells. Cardiac muscle thus forms a functional unit or syncytium. A contraction starting in one part of the heart spreads progressively throughout all the fibres. The electrical action potential of cardiac muscle is much longer than in skeletal muscle and persists throughout the contraction so that it has a correspondingly long refractory

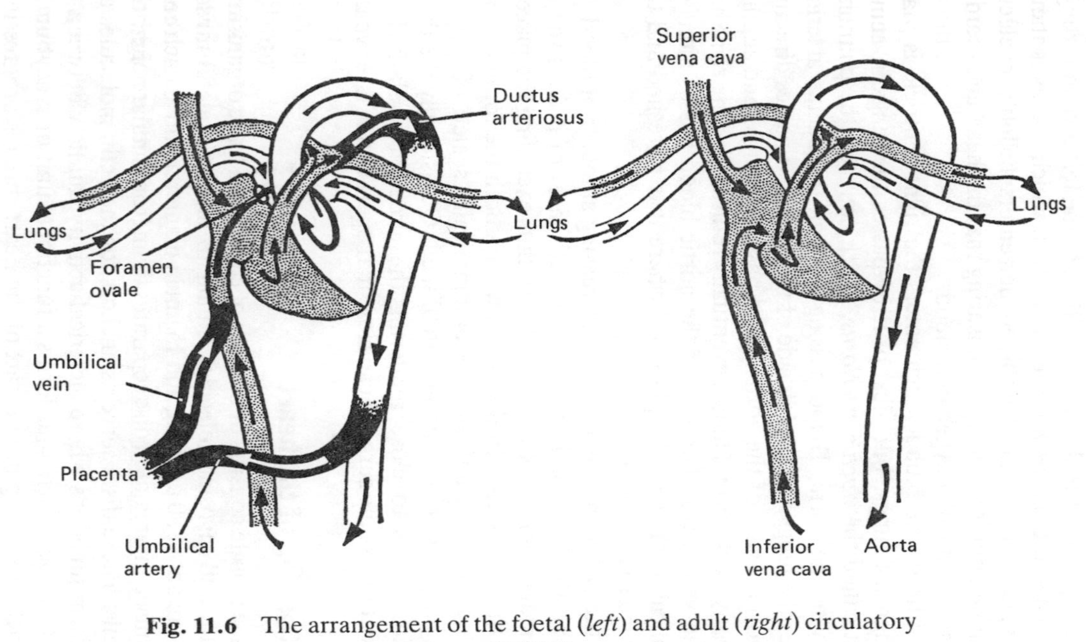

Fig. 11.6 The arrangement of the foetal (*left*) and adult (*right*) circulatory systems

period. Cardiac muscle cells are well supplied with glycogen and oxidative enzymes and are similar in this respect to the slow fibres of skeletal muscle (p. 47).

Heart muscle has a strong tendency to regular spontaneous contraction, although its beat is normally controlled by electrical signals from the *sino-atrial (SA) node*, or pacemaker, in accordance with the needs of the body. The SA node is situated in the right atrium and is regulated by the central nervous system via the autonomic nerve fibres. The electrical signal from the pacemaker first stimulates the walls of the atria (Fig. 11.7), which contract and squeeze blood into the ventricles. A fraction of a second later the electrical signal reaches the ventricles, through a narrow band of modified muscle fibres called the Bundle of His, and causes them to contract and eject their blood into the pulmonary artery and aorta. The muscle of the atria and ventricles then relaxes in sequence and their cavities fill with blood again. This cycle is normally repeated about seventy times a minute. The electrical activity of the heart may be recorded with an electrocardiograph. A typical record (Fig. 11.8) shows a P wave of atrial activity, a QRS complex of ventricular electrical activity and a T wave, which coincides with recovery to the resting state.

The cardiac cycle

The heart beats rhythmically but the rate varies with respiration. Atrial contraction is followed by ventricular contraction and the two chambers then relax in turn until the next contraction occurs. During the period of relaxation (*diastole*) the atria and ventricles fill passively with blood entering from the veins. The mitral or atrioventricular valve is wide open, but the aortic valve is shut. Towards the end of diastole the atrium contracts and forces or packs the last portion of blood into the ventricle. This event is signalled by the P wave in the electrocardiogram (ECG). About one-fifth of a second later the ventricle begins to contract and, as the pressure in the ventricle rises, the one-way atrioventricular valve snaps shut. The beginning of ventricular *systole* is signalled by the Q wave in the ECG. As the ventricle continues to contract the pressure inside it rises until it is higher than that in the aorta, the aortic valve is pushed open and the blood rushes into the aorta.

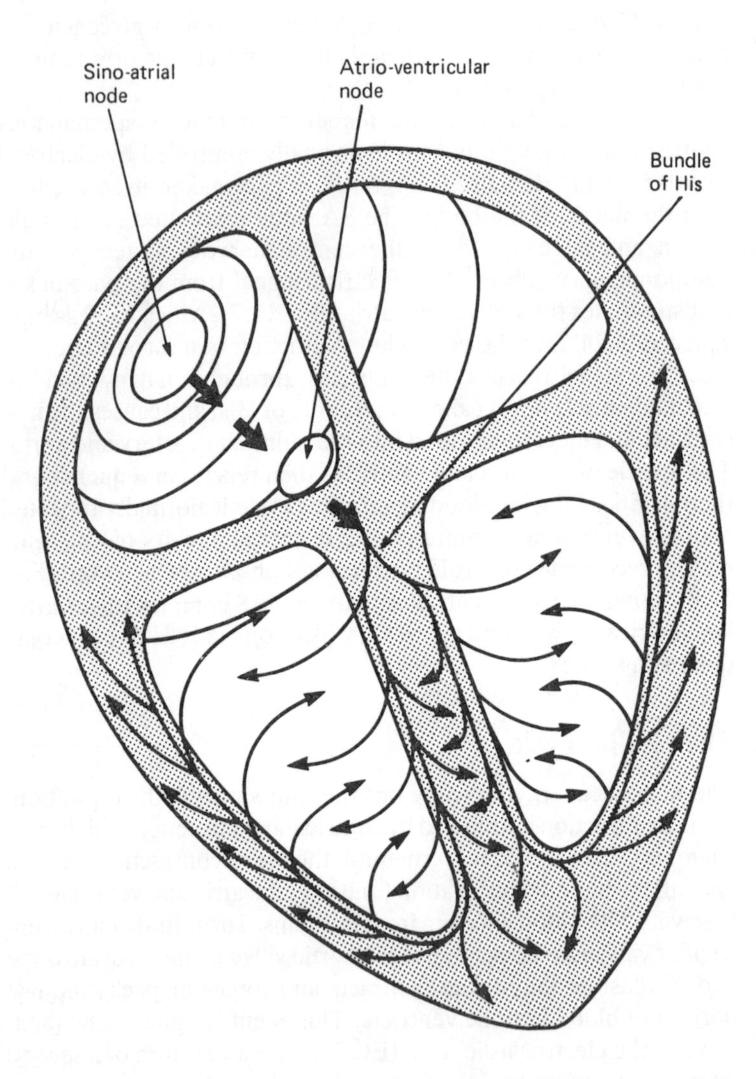

Fig. 11.7 Diagram of the initiation and spread of electrical excitation of the heart

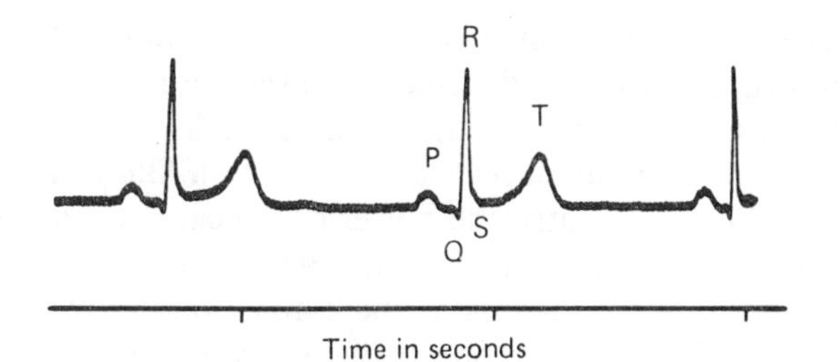

Fig. 11.8 Normal electrocardiogram. P represents atrial systole. QRST represents ventricular activity. Time: seconds. (See text)

The ventricle now relaxes and the pressure inside it falls. When pressure in the ventricle is lower than that in the aorta the aortic valve snaps shut. The ventricular pressure then falls very rapidly until it is lower than the pressure inside the atrium. The atrio-ventricular valve then opens and blood flows from the atrium into the ventricle. The opening and closing of the heart valves is an automatic process depending on the changes in pressure in front and behind each one. The complete cycle of events recurs a few moments later when the atrium contracts.

The duration of systole varies very little and slowing or speeding of the heart are largely accomplished by shortening or lengthening the diastolic interval between beats. Systole and diastole follow each other about once per second throughout the life of the individual, forming a regular sequence of events called the cardiac cycle.

Cardiac output

Experiments have shown that muscle fibres contract more strongly from a stretched position than when relaxed. The degree of stretch of the ventricular muscle fibres at the beginning of systole depends upon the amount of blood which has been packed into the ventricle during diastole. The output of the heart per beat (stroke volume) is directly proportional to the venous filling or cardiac input although the ventricles are never emptied completely. Normally the amount of blood ejected during each ventricular systole is approximately 60 ml per ventricle and the total cardiac output is about 3.5 to 5.0 litres per minute at rest. If the venous return is adequate cardiac output increases with the rate of the heart up to about 180 beats per minute but at faster heart rates diastole is too short for adequate filling of the ventricles and the output per minute falls. In addition to the variations in cardiac output dependent upon filling and rate, it is probable that the contractile power of the cardiac muscle fibres is increased by the action of adrenaline and noradrenaline. Sympathetic nervous activity increases the cardiac output by making the heart beat faster and more strongly. Parasympathetic nervous activity has the opposite effect. The factors of venous filling, nervous influences and hormonal activity work harmoniously together to control the response of the normal heart to the demands of the circulation and, ultimately, of every cell of the body.

Cardiac output varies constantly throughout life in response to a variety of influences. On standing up cardiac output falls by about a third, because gravity reduces the venous filling of the heart and the pressure in the atrium. Eating can increase the cardiac output by more than a litre per minute and excitement may increase it by a much greater amount. The maximal cardiac output of a man in a sedentary occupation may be only 15 litres per minute, but a trained athlete can achieve over 30 litres per minute. The heart of the athlete, and its stroke output, are larger and so for a given heart rate the quantity of blood ejected is greater.

Circulatory adaptations

The mechanisms just described are concerned in the adjustment of the circulation to various external influences.

Postural change

On first standing up from the lying position, blood is pooled in the veins and capillaries of the legs. This reduces the amount of blood returning to the heart and causes a lowering of the cardiac output and a fall in the arterial pressure. The amount involved is of the order of 500 cm^3. The fall in blood pressure evokes reflex responses from the carotid sinus (p. 110) which tend to oppose it through increased sympathetic nervous activity. This produces vasoconstriction of the arterioles and venules of the lower part of the body. The heart rate increases and the blood pressure rises again. This process is relatively slow and takes several seconds to complete. It is quite common to feel dizzy and for vision to become dim on standing. In extreme cases the reduction in cerebral blood flow may lead to fainting with loss of consciousness. These reflex responses fail rapidly with disuse, which can cause problems after space flights or immobilisation in bed by illness.

Haemorrhage

Sudden, severe haemorrhage produces a rapid fall in blood volume with a profound reduction in blood pressure. This is partially compensated for by a burst of sympathetic nerve impulses which reduces the overall capacity of the circulation by vasoconstriction. At the same time the heart rate increases and there is often sweating. The cardiac output falls and further vasoconstriction directs the remaining blood to the vital organs such as the brain and kidneys. In very severe cases consciousness may be lost or the kidneys may be permanently damaged.

Circulation through special areas

The brain

The cerebral blood flow depends mainly on the systemic arterial pressure and sympathetic nerve impulses cause little variation in the calibre of the blood vessels of the brain. When the circulation is failing arteriolar constriction causes a reduction in the blood flow through the internal organs, skin and muscles, while that through the brain and coronary arteries is relatively well maintained. However, if the heart stops for more than five seconds the victim loses consciousness and stoppage for more than fifteen seconds may result in epileptic fits or permanent damage.

Heart muscle

The muscle of the heart does not obtain its nutrition and oxygen directly from the blood passing through its chambers. The heart is supplied by the coronary arteries, which arise from the aorta close to the aortic valve. The blood flow through the coronary arteries depends primarily on the pressure in the aorta and upon the length of diastole. It is increased in conditions in which the sympathetic nervous system is overactive.

The skin

The blood flow through the skin is controlled by the sympathetic nervous system. This is of particular importance in the control of heat loss by conduction and radiation from the skin (pp. 66, 177). Exposure to warmth leads to reduced sympathetic nervous activity and to vasodilatation with flushing and loss of heat through the skin.

Control of the circulation

The heart and blood vessels form a closed system with smooth, elastic walls. Each heart beat pumps blood into this system so that the blood is always under pressure but constantly moving onwards. The walls of the aorta and larger arteries are distensible and resilient because they consist largely of elastic tissues with very little muscle. The walls of the smaller arteries, and particularly of the arterioles, contain relatively larger amounts of muscle. The coordinated relaxation and contraction of the muscles of the walls of the smaller arteries and arterioles varies their resistance to the passage of blood. This resistance determines the proportion of the heart's output that will flow into any organ or part according to need. The distribution of blood to various parts of the body depends on the degree of contraction of the smaller muscular arteries, acting like taps. By the time the capillaries are reached the friction of the continually narrowing arteries reduces this pressure to about one-quarter of its level at the heart. There is a further drop in pressure in the capillaries. Consequently, the pressure in the veins is very low and by the time the atrium is reached it is negligible. Blood returns uphill from the legs only because of the non-return valves in the veins and the continuous pumping action of the limb muscles in walking.

The distribution of the blood in the body depends upon the rate and force of the heart beat and the opening and closing of the small arteries. The oxygen needs of the different organs vary in relation to their activity, increasing when they are working. The blood flow to the organs is controlled by the size of the small arteries leading to them. At any given time the plain muscle in the majority of the small arteries is contracted, directing the blood into the active organs and away from the inactive ones. If all the arterial muscle relaxed at the same time the capacity of the circulation would much increase. The circulation would fail because there would be far too little blood to fill the dilated blood vessels and maintain the correct pressure.

Nerve signals from the blood vessels entering and leaving the heart and chemical signals carried in the blood converge on the base of the brain (Fig. 11.9). They evoke further nerve signals, which travel in the fibres of the autonomic nervous system to the heart and arteries. Chemical stimuli locally in the tissues dilate or contract the blood vessels and thus change the peripheral distribution of blood. These mechanisms work together to produce a coordinated, delicately adjusted circulation of blood exactly suited to the needs of the body. When resting, the pulse is slow, the heart output is low and the blood pressure is minimal. Activity leads to a faster pulse rate, larger heart output and higher blood pressure. Most of the blood is now directed to the active muscles by contraction of the 'taps' formed by the small arteries in other parts of the body. The capillaries of the muscles are widened by the direct action of carbon dioxide and waste products of metabolism on their walls.

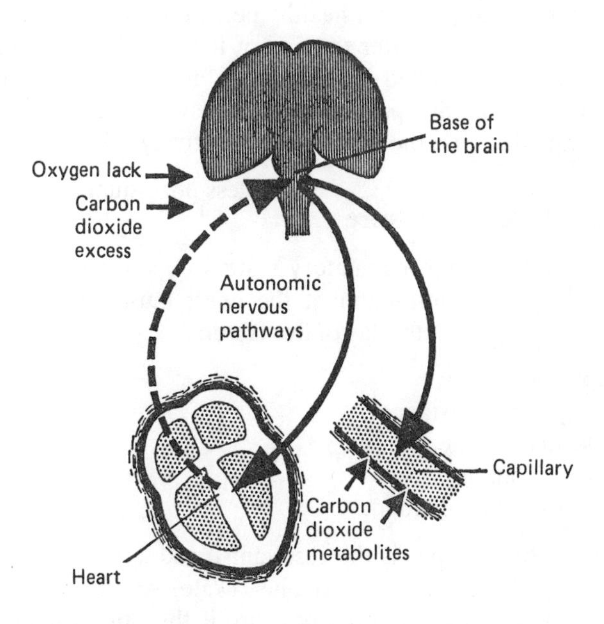

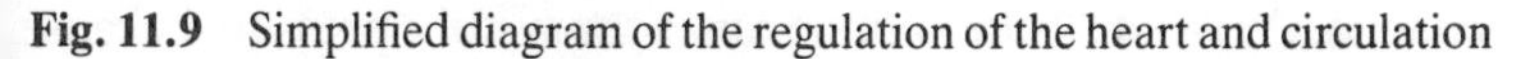
Fig. 11.9 Simplified diagram of the regulation of the heart and circulation

The pulse

The ventricles contract, expelling the blood they contain, and relax to refill passively from the veins, about once a second. This is the cardiac cycle, which is divided into contraction or *systole* and relaxation or *diastole*. Each contraction squirts 60 ml (2 fluid oz) of blood out of each ventricle. Every heart beat thus raises the pressure for a moment throughout the arterial system of the body. This can be seen and felt as the arterial pulse in, for example, the radial artery at the wrist. This shock wave takes a little time to travel from the heart to the wrist, so that the radial pulse occurs after, and not simultaneously with, cardiac systole, and is always felt slightly after the apex beat. The normal pulse rate is about 70 per minute.

Many factors influence the resting pulse rate. It tends to rise with age and there is also a constitutional factor. Athletes, for example, often have slower pulse rates than the non-athletic. In the healthy person the resting pulse rate may be anywhere between 50 and 80 beats per minute. Exercise increases the pulse rate up to nearly 200 beats per minute, but in a healthy person it returns to the resting value a few minutes after the end of the exercise. In general the pulse rate is controlled by changes in activity of the autonomic nervous system. Sympathetic and parasympathetic impulses increase and decrease the heart rate respectively.

Arterial blood pressure

The pressure in the circulatory system depends upon the force exerted by the contraction of the heart muscle but its level is maintained by the interplay of five factors:

1 Cardiac output;
2 Peripheral resistance;
3 Elasticity of the arteries;
4 Blood volume;
5 Viscosity of the blood.

The cardiac output is the resultant (product) of the volume of blood expelled per beat and the heart rate. An increase in cardiac output increases the arterial pressure if the other factors do not change. A rise in stroke volume raises the systolic pressure because a larger amount of blood has to be accommodated in the arterial system. Speeding the heart tends to increase the diastolic pressure more than the systolic because the time between beats (diastolic interval) is less and there is less time for the diastolic pressure to fall. Arterial blood pressure is usually measured in mm of mercury (Hg). The normal range is 120/70 mm Hg.

The peripheral resistance is directly related to the calibre of the arterioles, which is controlled by both nervous and chemical factors. Narrowing of the arterioles, the cardiac output being maintained, leads to a rise in arterial pressure until the outflow through the narrowed arterioles is again equal to the input from the heart.

The elasticity of the aorta and large arteries, in conjunction with the peripheral resistance, modifies the arterial pressure and transforms the intermittent pressure pulses imparted to the blood by the heart beat into a steady stream from the far ends of the arterioles. If there were no elasticity in the system the systolic pressure would be much higher, but the diastolic pressure would tend to be lower because only the peripheral resistance would be maintaining it.

The walls of the arterial system are normally under tension, or stretched, so that the system is, in effect, always overfilled with blood. If the blood volume is decreased by haemorrhage or other form of fluid loss the blood pressure falls. The arterioles constrict to counteract this, but if their reaction is inadequate both the systolic and diastolic pressures fall.

The viscosity of the blood is also a factor in maintaining the blood pressure; if the circulatory system were filled with treacle the pressure would evidently have to be much higher to maintain the normal rates of flow. In practice, the viscosity of the blood is of much less importance in modifying the blood pressure than is the peripheral resistance of the arterioles.

In summary, the arterial blood pressure depends primarily upon two factors, the vascular resistance and the cardiac output, but it is also affected by the elasticity of the large arteries and by changes in the volume and viscosity of the blood. The pressure present in any artery between the pulse waves is called diastolic. The peak of the rise which occurs after every ejection of blood into the aorta is the systolic pressure.

Effects of exercise

At complete physical rest the cardiac output is small but the healthy body responds to any physical or emotional demands by increasing the cardiac output, raising the arterial pressure and augmenting the circulation of the blood. Before exercise is undertaken the brain prepares the body by liberating adrenaline from the adrenal glands and increasing sympathetic vasoconstrictor activity. This has the effect of increasing the heart rate and cardiac output, thus raising the arterial pressure. When the exercise begins, these processes continue and are augmented by the accumulation of metabolites resulting from the muscular contraction. The effective return to the heart of the larger quantities of blood flowing through the muscles is ensured by the pumping action of respiration and of the muscles themselves. This leads to an increase in the cardiac output and in the circulation through the muscles, which enables adequate supplies of oxygen to reach them and ensures the removal of waste products. In sustained physical exertion these mechanisms may be inadequate to supply enough oxygen and the muscles may then build up an 'oxygen debt' (p. 48).

There are also controls which prevent these responses becoming excessive. If exercise or other influences tend to cause an undue increase in blood pressure, the vasomotor reflexes come into action. Nerve endings which are sensitive to arterial pressure are found in the aorta and the internal carotid arteries on both sides. These nerve endings respond to increase arterial pressure by sending nerve impulses up to the medulla, where they evoke a reflex slowing of the heart and a fall in the blood pressure. This is achieved by a reduction in the normal outflow of sympathetic nerve impulses. These reflexes are constantly in action controlling and smoothing out the response of the heart and circulation to the varying conditions of existence. In addition to these nervous controls, adrenaline and noradrenaline secreted by the suprarenal glands also influence the circulation; their overall effect is to increase arterial blood pressure.

12

The Nervous System

The nervous system is the most elaborate and complicated part of the body. Its function is to coordinate and control all the multifarious activities of the organism; survival of the body without the nervous system would be impossible. Rapid communication between the various parts, the effective, integrated activity of different organs and tissues and the coordinated contraction of muscle are almost entirely dependent upon it. It consists of billions of excitable cells – *neurones* – arranged in intricate linked networks. They are surrounded, supported and nourished by a multitude of non-excitable cells – *neuroglia*. Neuroglia take the place of the tissue fluid found elsewhere in the body and transport gases, nutrients and waste products between the neurones and the blood vessels.

To perform its innumerable and essential functions the nervous system consists of three fundamental parts, the *receptors*, the *central neuraxis* and the *effectors*. Receptors, which respond to changes in the environment or in the body itself, are connected by sensory nerve fibres to the central mass of nerve cells or neurones. This is the sensory system, which furnishes information essential to the economy and survival of the body. The constant, ever varying flood of information passes into the central nervous system where it gives rise not only to effective, coordinated responses in muscles, glands and blood vessels but to the remarkable and unexplained phenomena of consciousness, vision, sound, memory, thought and emotion. Finally, there is the effector system of motor nerve fibres through which are conveyed the nerve impulses which finally result in the manifest responses to the input of sensation.

Neurones receive, transform, integrate, encode and transmit information. There are about 100 billion of them in innumerable different forms in the adult human brain (Fig. 12.1). All the space between the neurones is filled by neuroglial cells, which number five or even ten times as many as the neurones. Each neurone has a cell body containing its controlling nucleus and all the intricate mechanisms for carrying out its biochemical processes. Each neurone has a bushy branching system of delicate tubular processes – *dendrites* – making up to 80 per cent of the neuronal surface. The dendrites are covered by contact points – *synapses* – derived from other neurones. Each neurone may receive 1000 to 10 000 synapses from up to 1000 other neurones. There may be 100 trillion synapses in a complete brain. Each synapse contains vesicles holding several thousand molecules of chemical transmitter. In contrast to the peripheral nervous system there are more than 20 known transmitter substances in the central neuraxis.

Each neurone has a single long *axon* leading away from it and signal transmission is always one way: dendrite → cell body → axon

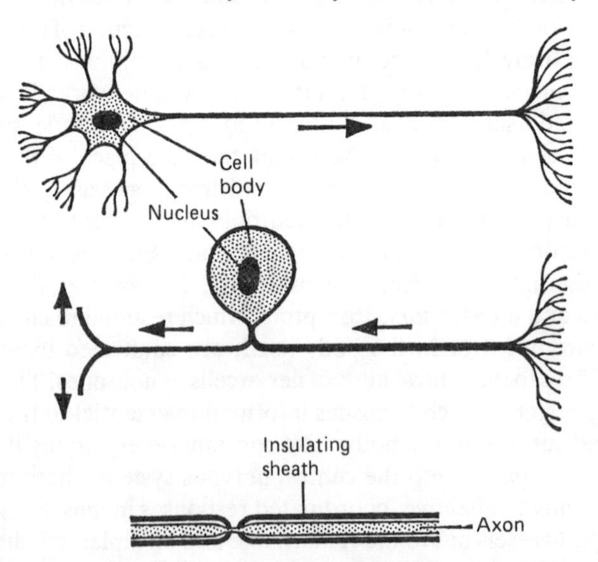

Fig. 12.1 Types of nerve cells (not to scale). *Upper:* Unipolar cells with dendrites on the left and axon travelling to right. *Middle:* Bipolar sensory cell leading from receptors on the right to the spinal cord on the left. *Lower:* Longitudinal section of axon, highly magnified

→ nerve ending. Some neurones carrying sensory signals to the spinal cord have two axons and no dendrites. The axons or nerve fibres ramify throughout the body and may be up to 1 metre long. They are usually covered with insulating sheaths of fat and protein – *myelin* – which prevents interference between the electrical signals of neighbouring neurones. Nerve fibres vary in diameter from 20 μm to fractions of a μm. The differences in diameter depend largely on differences in thickness of the myelin sheaths. The myelin is formed by the Schwann cells which enclose the peripheral nerve fibres.

Neuroglia are supporting cells. Some, the *microglia*, enter from the capillary blood vessels and act as scavenger cells. The larger *oligodendroglia* have few processes and are engaged in myelin formation. The smaller *astrocytes*, with numerous processes, have a metabolic function.

The blood-brain barrier

Nerve cells are very sensitive to the concentration of the substances around them and the blood-brain barrier is responsible for maintaining the environment necessary for normal neuronal function. This barrier is apparently situated at the endothelial lining of the capillary blood vessels and seems to be related to the size of the molecules concerned. One effect is that many substances such as antibiotics and anaesthetics cannot enter the brain from the blood vessels and this may interfere with the treatment of illness.

The nerve impulse

In most cells there is a difference in electrical potential across the cell membrane and in neurones the interior is −70 to −90 mV to the outside. This resting potential is chiefly due to K^+ diffusing out of the cell down its large concentration gradient. The outward movement of K^+ ions carrying positive charges builds up a membrane potential which opposes and eventually balances out further diffusion. A similar equilibrium potential also develops for other ions and the exact value of the membrane potential depends upon the relative values of the various ionic permeabilities, especially for sodium and potassium. In nerve and muscle the *resting potential* is close to the potassium equilibrium potential but in erythrocytes, for example, it is about midway between the sodium and potassium values.

The resting potential fluctuates gently because of ion movements. When it increases the neurone is less excitable and when it decreases the neurone is more excitable. A nerve impulse consists of a sudden change in the resting potential of the neurone and it is accompanied by an *action potential*. At its start there is a rush of Na^+ ions through the membrane which lasts for about a millisecond and the membrane potential becomes positive. After the nerve impulse has moved on, the membrane potential is quickly restored to its normal negativity by the continuing leak of K^+ ions from the cell and the activity of the Na^+ pump (p. 15). The action potential sets up a flow of ions through the neighbouring membrane and the resulting depolarisation (voltage reversal) spreads producing a propagated action potential. The speed of a nerve impulse is related to the thickness of the fibre carrying it: the larger the fibre the faster the impulse, because the nodes are further apart. Fibres with a diameter of 20 μm transmit nerve impulses at over 100 metres per second, which is ten times faster than the fastest sprinter, but in the smallest fibres impulses travel at less than walking pace.

Recovery after an action potential has been generated takes about 5 milliseconds and the membrane then behaves normally again. During this period – the *refractory period* – the neurone is absolutely or partially incapable of generating another action potential. Each action potential is identical and is generated at a certain stimulus threshold – the all or none response. Fluctuations in membrane potential below the firing level have no effect. In most neurones the upper limit of rate of firing is about 100 per second but some specialised cells in the spinal cord can reach about 1000 per second. The energy for both transmission and recovery is provided by oxidative processes. Neurones die in a few minutes if they are deprived of oxygen, but they also die quickly if they are not supplied with enough glucose.

All axons end on another neurone, muscle cell or some similar element by means of specialised endings – synapses. The billions of neurones are anatomically separate units with no protoplasmic continuity. Conduction of impulses in the central nervous system is normally one way from dendrite to cell body to axon. The synapses are structurally distinct, expanded portions of the axons and they are very closely applied to the dendrites or cell body of the neighbouring neurone, but the two always remain anatomically distinct. Each neurone has many synapses in contact with it and is directly or indirectly connected with neurones throughout the nervous system.

Most synapses are excitatory in action and when a nerve impulse reaches a synapse it releases a chemical transmitter such as noradrenaline which tends to fire off a response in the neighbouring neurone. Some synapses, however, release a different chemical, such as γ-aminobutyric acid, which opposes the tendency to firing, i.e. they are inhibitory. The exact moment of firing of a neurone depends on the interplay and balance of excitatory and inhibitory synapses on its cell body and dendrites, and upon the glucose and electrolyte content and temperature of the tissue fluid bathing it.

The peripheral nervous system

Throughout life there is a constant traffic of nerve impulses travelling into and out of the nervous system. In addition to the eyes and ears, which respond to light or sound waves only, the whole body is provided with innumerable nerve endings which respond to changes in the energy impinging on them and to changes in the state of the structures of the body itself. The richest variety of nerve endings is found in the skin. They vary from elaborate interlacing networks of free nerve filaments to round discs and minute basket-like coils. The free nerve endings are generally believed to respond to energy levels that may be dangerous to the body. They set up signal patterns which pass to the brain and give rise to the sensation of pain. The more elaborate receptors are thought to be responsible for the nerve signals which underly sensations such as touch, warmth and pressure. These ingoing or *afferent* nerve impulses travel in sensory nerve fibres to the spinal cord.

The outgoing or *efferent* nerve impulses traverse motor nerve fibres and end in the muscles. These impulses are initiated by nerve cells in the anterior horns of the spinal cord or in special nuclei in the brain stem, which are themselves activated by nerve impulses playing upon them from other parts of the nervous system. The motor and sensory nerve fibres enter and leave the spinal cord by separate channels but they soon mingle inextricably in the peripheral nerves which supply all parts of the body. With a few exceptions all the peripheral nerves contain both motor and sensory nerve fibres and are called mixed nerves. They also contain many fibres belonging to the autonomic nervous system.

In addition to the impulse patterns evoking conscious sensations, the central nervous system is continually bombarded with signals from the interior of the body which may never reach consciousness. These signals result from changes in pressure, temperature, blood circulation, muscle length and loading, and from many other changes in the body. They form the basis of the innumerable self-regulating mechanisms that work together to maintain the health and integrity of the body.

The peripheral nervous system consists of twelve pairs of *cranial nerves* arising from the brain or brain-stem and thirty-one pairs of *spinal nerves*. The cranial nerves consist of both motor and sensory nerve fibres except for the first, second and eighth which are purely sensory. All the spinal nerves contain both motor and sensory fibres. The cell bodies of the spinal motor nerves are situated in the anterior horns of the spinal cord. The motor nerve fibres branch repeatedly and each branch ends on a different muscle fibre forming a motor unit (p. 51). The cell bodies of the sensory nerve fibres are located in *ganglia* on the posterior nerve roots outside the spinal cord.

The spinal nerves from the eight cervical and first thoracic segments of the spinal cord supply the motor and sensory fibres for the neck and upper limbs. On leaving the vertebral column they form an intricate network of branches comprising the cervical and brachial plexuses. The brachial plexus gives rise to the three large mixed nerves of the upper limb – median, ulnar and radial.

The thoracic spinal nerves with the first lumbar nerve supply the sensory and motor nerve endings of the chest and abdomen. The lower four lumbar and the sacral nerves form the big lumbosacral plexus which serves the whole of the lower limb via the sciatic and femoral nerves.

The central neuraxis

With the exception of the sensory neurones, which have their cell bodies in the posterior nerve root ganglia, and parts of the autonomic nervous system (Chapter 13), all the nerve cells are situated in the central neuraxis. This consists of the spinal cord, brain stem and the cerebral hemispheres. The central neuraxis is composed of billions of cells, arranged in sheets and masses, called the *grey matter*, and an immensely complex interlacing network of nerve fibres, the *white matter*.

Neurones with comparable functions tend to be located near each other, forming functionally similar areas and masses. The fibres linking them with the rest of the neuraxis are usually arranged in separate, identifiable cords or trunks, often called *tracts*.

The spinal cord

The spinal cord lies in the vertebral canal and consists of the central H-shaped grey matter, surrounded by white matter (Fig. 12.2). The white matter is composed of myelinated nerve fibres, which run up and down the spinal cord linking together the various parts of the nervous system. The sensory fibres enter via the posterior nerve roots and form synaptic connections with other neurones. Fibres from these secondary sensory neurones cross over to the opposite side and travel to the cerebrum through the posterior columns and the spinothalamic tracts. Others pass up the spinocerebellar tracts to the cerebellum on the same side. Motor nerve fibres descend in the pyramidal tracts, crossing to the opposite side in the brain stem and spinal cord, and terminate in the anterior horns of grey matter which supply the nerve fibres emerging in the anterior nerve roots on their way to the various muscles.

The spinal cord gives rise on each side to 31 pairs of nerve roots anterior and posterior, which join together to leave the vertebral canal, through the corresponding intervertebral foraminae. There are 8 cervical, 12 thoracic, 5 lumbar, 5 sacral and 1 coccygeal pairs of nerve roots on each side. Each anterior root consists entirely of outgoing or efferent fibres and is motor in function, while each posterior root consists of efferent or ingoing fibres and is sensory in function.

The brain

The sensory signals arising in all parts of the body travel along nerve fibres to the spinal cord and ascend to the brain. The brain has three main parts (Fig. 12.3) – cerebrum, cerebellum and brain stem. The

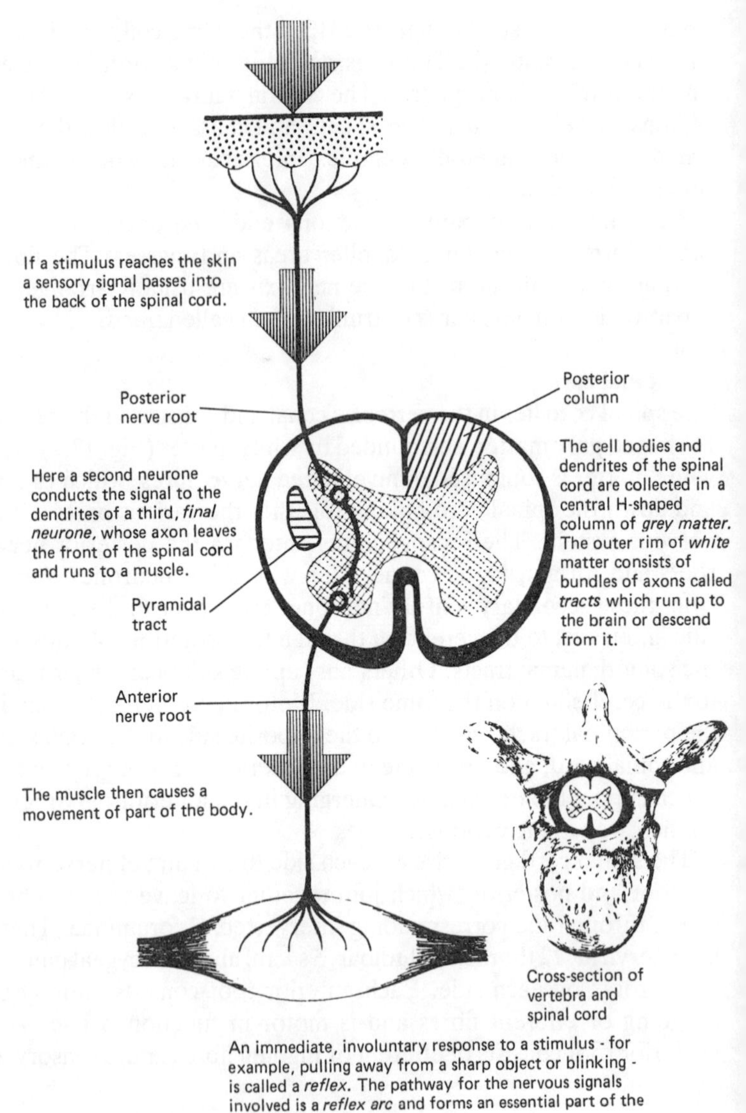

Fig. 12.2 Reflex pathway via the spinal cord

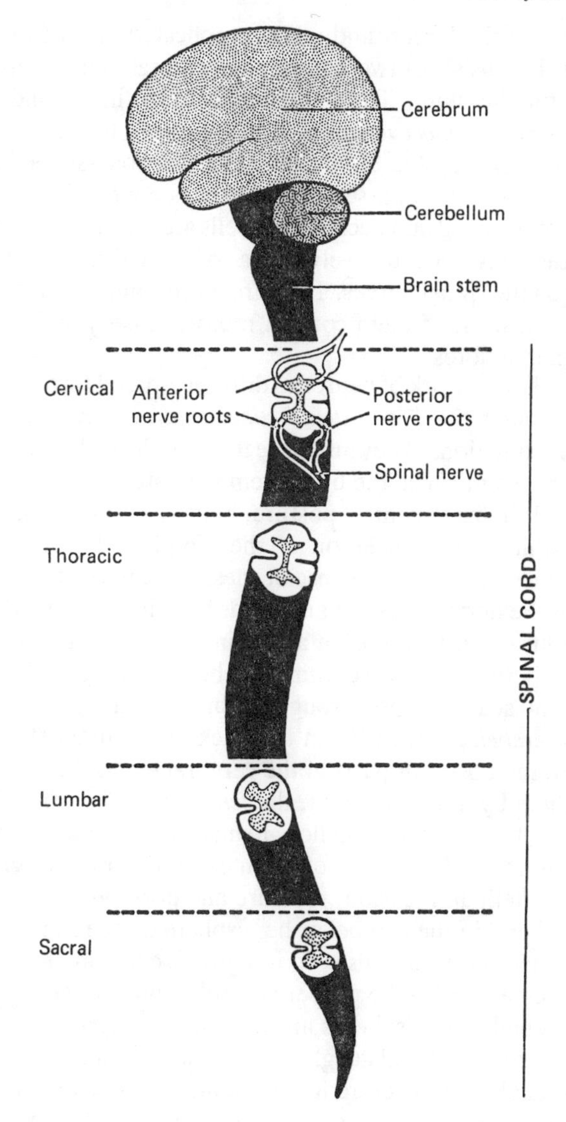

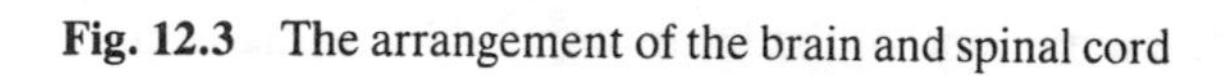

Fig. 12.3 The arrangement of the brain and spinal cord

cerebrum is the largest and most complicated part of the nervous system. It consists of two halves, the cerebral hemispheres, which nearly fill the skull. They are linked by a thick band of white matter – the *corpus callosum* – and are joined to the upper part of the spinal cord by the *brain stem*. This is a cone-shaped structure which contains the masses of nerve cells controlling the various nerves of the head and neck. These cells are arranged in identifiable nuclei and give rise to twelve pairs of cranial nerves, which are similar to the spinal nerves. They are more complex than the spinal nerves but most of them contain motor, sensory and some autonomic nerve fibres.

There is a network of neurones diffusely spread through the brain stem forming the *brain stem reticular formation*. Its cells have manifold functions. They influence the overall level of activity of the nervous system which the brain stem regulates, restrains and coordinates. It helps to control posture, balance and the vital functions of breathing and circulation of the blood. It is concerned with wakefulness and level of consciousness. It receives signals from all parts of the nervous system and sends back impulses modifying the behaviour of both cerebrum and spinal cord. All the nerve fibres running between the cerebrum, cerebellum and spinal cord, both motor and sensory, pass through the brain stem.

The *cerebellum* arises from the back of the brain stem and lies beneath the posterior parts of the cerebral hemispheres, separated from them by a fibrous sheet, the *tentorium*. The cerebellum also consists of two, much smaller, hemispheres. The sensory nerve impulses from the organs of balance and those concerned with muscle length and tension, pressure and position of the bones and joints all end in the cerebellar hemisphere on the same side of the body. These nerve impulses do not give rise to conscious sensation but are essential for the proper control of posture and movement. The cerebellar hemispheres are in turn connected to the cerebral hemispheres, the basal ganglia, the brain stem and the spinal cord. The cerebellum is a centre for the control and coordination of the posture, balance and movement of the whole of the body under all circumstances. It both strengthens and modifies reflex responses and movements. It provides the muscular background and postural basis for all forms of muscular activity.

At the upper end of the brain stem in the centre of the brain lies the *hypothalamus*. It is a small collection of nuclei and fibres closely related to the pituitary gland which it largely influences. It is essential for the regulation and control of visceral activity, body temperature, water and electrolyte balance, blood pressure, sexual and reproductive activity and possibly body weight.

Cerebral hemispheres

Each cerebral hemisphere is basically responsible for sensation and movement of the opposite half of the body but some automatic movements such as breathing and swallowing have dual control from both sides and are not abolished if one hemisphere is damaged.

During development powers of speech develop and one hand becomes predominant for skilled movements such as writing, throwing and cutting. In most people this is the right hand, which is controlled by the left cerebral hemisphere. This hemisphere also controls speech, reading, writing and special skills. It is called the dominant hemisphere. In left-handed people the right hemisphere controls speech, reading and writing.

Each cerebral hemisphere contains a large mass of nerve cells – the *optic thalamus* – in which the bulk of the sensory signals from the opposite side of the body relay. These sensory signals are transmitted from the thalamus to the thick mantle of nerve cells – the grey matter or *cerebral cortex* – which enfolds each cerebral hemisphere. Each hemisphere thus deals mainly with signals originating from stimuli involving the opposite side of the body. Deep in each cerebral hemisphere, in addition to the thalamus, lie the *basal ganglia*. These are masses of nerve cell bodies arranged in three main blocks – the caudate nucleus, putamen and globus pallidus. They are connected to each other, to the cerebrum and to the cerebellum by a network of nerve fibres. They send nerve fibres downwards to connect up with other cell bodies in the brain stem and spinal cord. The basal ganglia, with all their nerve fibre connections, form the extrapyramidal nervous system, which regulates the postural activity of the muscles of the body. It is an essential foundation for the performance of the ceaseless, habitual, automatic movements underlying the smooth performance of the skilled, purposive activities of every day life (Fig. 12.4).

The grey matter contains the cell bodies of the motor neurones,

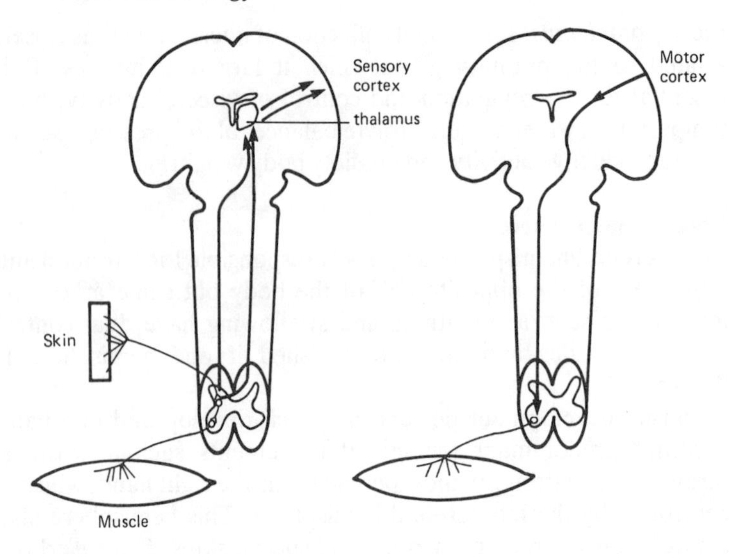

Fig. 12.4 Sensory and motor pathways in the brain and spinal cord

which run down the brain stem and spinal cord, crossing to the other side as they descend. They initiate and conduct the patterns of motor nerve signals which travel down to the spinal cord and eventually produce voluntary muscular contraction on the opposite side of the body. The final link in the pathway is a nerve cell in the spinal cord which sends a nerve fibre out to a particular muscle and controls the activity of a single motor unit (p. 51) within it. Any effective muscular movement results from the organised, linked activity of large numbers of nerve cells in the cerebrum, the cerebellum and the brain stem, producing coordinated, precise contractions of muscles in the trunk and limbs together. Even a simple action like walking is a series of events involving nearly every muscle in the body.

The higher, and mysterious, functions such as thought, memory and consciousness are properties of the brain as a whole rather than of any particular part of it. The understanding and production of speech, reading and writing are dependent in right-handed people upon the integrity of the left cerebral hemisphere only. Certain other areas of the cerebral hemisphere are of special importance for different activities (Fig. 12.5). Self-control and co-operative,

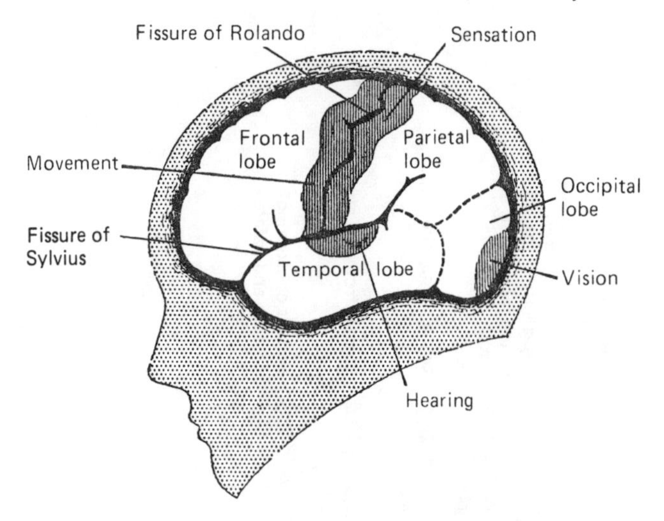

Fig. 12.5 Diagram showing the conventional plan of the localisation of function in the cerebral hemisphere

socially desirable behaviour seem to be related to the proper functioning of the *frontal lobes* of the brain. They extend forward from the *fissure of Rolando* and also contain the nerve cells which initiate voluntary movement.

Immediately behind the fissure of Rolando lie the *pariental lobes*. They receive and respond to sensory nerve impulses from all over the body. The detailed appreciation, assessment, integration and judgement of sensation depend on the parietal lobes. The left parietal lobe, in right-handed people, is concerned with the understanding and reception of speech, language and with reading.

The posterior parts of the cerebral hemisphere form the *occipital lobes*, which contain the nervous structures necessary for vision (Fig. 12.6). The neuronal equipment necessary for the recognition of light, shade, shape and colour is localised at the back and the more forwardly placed structures are responsible for the appreciation and interpretation of visual patterns.

The lower parts of the cerebral hemispheres jut out to form the *temporal lobes*, which are separated from the parietal and frontal lobes by the *fissure of Sylvius*. The temporal lobes are especially

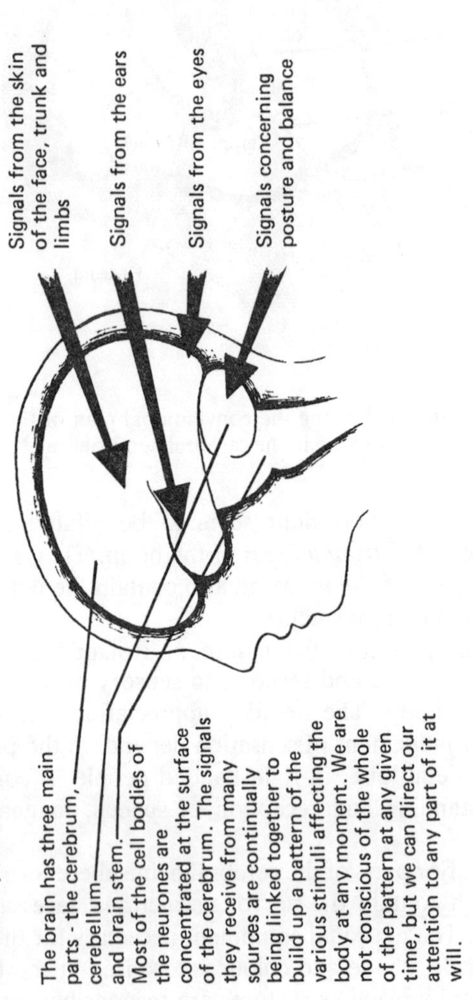

The sensory areas of the brain

The brain has three main parts - the cerebrum, cerebellum and brain stem. Most of the cell bodies of the neurones are concentrated at the surface of the cerebrum. The signals they receive from many sources are continually being linked together to build up a pattern of the various stimuli affecting the body at any moment. We are not conscious of the whole of the pattern at any given time, but we can direct our attention to any part of it at will.

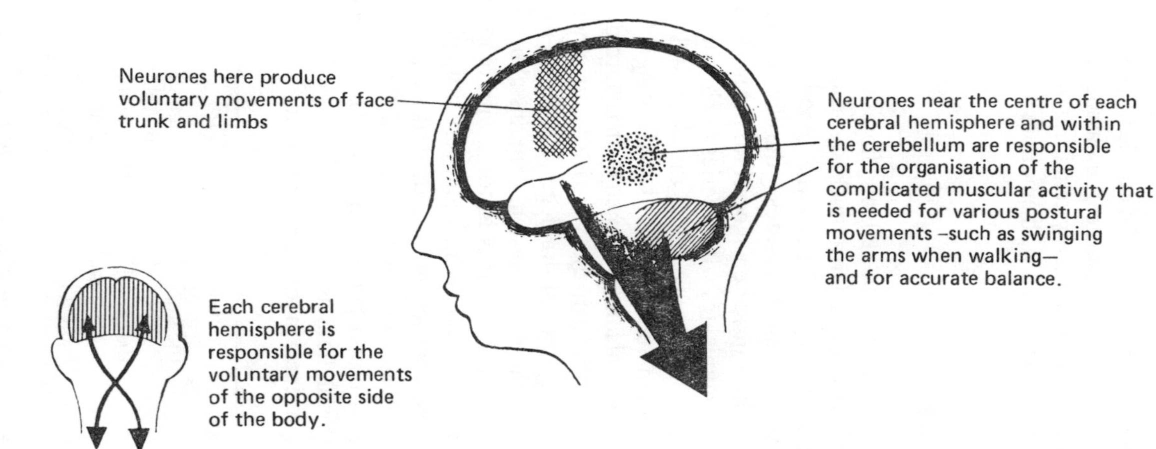

Fig. 12.6 The functional organisation of the brain

concerned with hearing but their other functions are poorly understood although they seem to be concerned to some extent with memory.

The *limbic system* is a strip of cortical tissue deeply situated at the root of each cerebral hemisphere plus certain associated nuclei of grey matter. The anterior part of this system is related to the sense of smell while the rest is concerned with emotions such as rage and fear or with sexual behaviour.

Cerebrospinal fluid

The enormous aggregation of nerve cells and fibres constituting the brain and spinal cord is suspended in a shock-absorbing liquid. This cerebrospinal fluid is filtered off from the blood plasma. The filtration occurs in two vascular structures – the *choroid plexuses* – which project into cavities inside each cerebral hemisphere. Each cavity or ventricle communicates with a series of channels leading eventually to the fluid-filled space lining the skull and vertebral column in which the brain and spinal cord are floating. Nearly 0.5 litre (1 pint) of cerebrospinal fluid is formed by the choroid plexuses every day. It circulates through the ventricles and other spaces before being absorbed by special drainage tufts at the top of the skull. At any one time the whole system contains about 190 cm^3 (1/3 pint) of cerebrospinal fluid. It is isotonic with blood plasma but contains less potassium and calcium and very little protein. Its sodium and chloride content are greater than plasma. Table 12.1 summarises the differences. The concentrations of glucose and urea are similar to those in blood plasma. In addition to mechanical protection cerebrospinal fluid is concerned in the transport of nutrients and waste products in the nervous system.

Substance	*Plasma*	*CSF*
Na	145 mmol/l	143 mmol/l
K	5 mmol/l	3 mmol/l
Cl	100 mmol/l	114 mmol/l
Ca	2.5 mmol/l	1.2 mmol/l
Protein	6.0 g/l	20 mg/l

Table 12.1 The composition of cerebrospinal fluid

Transmission of motor and sensory nerve impulses

The neurones responsible for voluntary movement are situated in the posterior part of the frontal lobe just in front of the fissure of Rolando. They are arranged in a strip running laterally from the mid-line. The neurones supplying the foot come first and the rest of the neurones are arranged in order with face outermost. There are about 30 000 large neurones (Betz cells) on each side. They give rise to large rapidly conducting nerve fibres which cross over in the brain stem and form part of the *pyramidal tract*, in which they pass down the spinal cord to end in the anterior horns all the way down. Here they form synapses on the anterior horn cells giving rise to the motor neurones, which form the final common path to the muscles themselves. The initiation and control of voluntary movement is described later.

Each posterior nerve root of the spinal cord contains sensory nerve fibres from a particular strip of the skin of the body or limbs, called a *dermatome*. The sensory nerve fibres travel inwards to the spinal cord and eventually terminate on the cell bodies of the second sensory neurones. The axons of these neurones then cross to the opposite side of the spinal cord or brain stem and end in the *optic thalamus* on that side. Here they give rise to impulses in the third sensory neurones, which end in the cortex of the *parietal lobe* of the cerebral cortex. The parts of the body are represented in a fixed pattern in the cortex of the parietal lobe just behind the fissure of Rolando. The foot area is located just at the mid-line and the leg, thigh and trunk areas spread out in a strip downwards towards the base of the skull. Then come the areas of the hand and arm, followed by the face and mouth, which lie at the bottom end of the Rolandic fissure. The parts of the body with the most precise sensory powers, such as the hand and face, are represented in larger areas of cortex than those with little discrimination, like the trunk. The arrangement is very similar to that just described for the motor cortex.

There is considerable reorganisation, rerouting and even suppression of sensory nerve impulses during their upward journey, largely under the influence of the brain stem reticular formation, so that essential information receives priority of transmission. The optic thalamus is of particular importance as a centre for integrating and correlating sensory impulses. It also subserves a level of crude sensory appreciation, possibly concerned with the experience of pain. Some of the nerve impulses from muscles, bones and joints travel along pathways which terminate in the cerebellar hemispheres of both the same and of the opposite sides.

The nerve impulses concerned in bodily sensations are directed to the parietal lobe, which is sometimes called the primary sensory area. Similarly the occipital lobe is the primary visual area and the primary areas for taste and smell are situated deep in the *Sylvian fissure* and in the *temporal lobe* respectively. The primary sensory areas are richly connected by nerve fibres to the rest of the cerebral cortex. The areas of cortex lying between the primary sensory areas are the *association areas*, and it seems to be here that the recognition and differentiation, the significance and meaning of sensation arise. In these areas, too, the necessary correlation of sensory impulses occur upon which purposive activity can be based.

There are several types of sensory receptor in the skin and deeper tissues, each responding preferentially to a particular stimulus. Mechanical deformation and chemical and temperature changes, set up movements of ions which change the resting membrane potential and establish a localised generator potential in proportion to the strength of the stimulus. The generator potential eventually evokes a series of action potentials in the sensory nerve fibre with a frequency related to the strength of the stimulus. If the stimulus persists the frequency falls and the receptor undergoes *adaptation*. This occurs with sensations other than pain. The initial response to a stimulus is followed by a progressive diminution in sensation. The heat of a hot bath rapidly changes to a pleasant warmth and the contact of clothing is only felt intermittently when movement changes the pressure between fabric and skin. In addition, the pattern of activity in the afferent pathways is continually monitored and modified by nerve impulses in fibres passing down from higher levels in the nervous system. These influences operate at both spinal and brain-stem levels, further changing the final sensation.

Each sensory nerve fibre has a number of endings forming a *receptive field*. The fields from different nerve endings overlap. The variation in sensitivity between various areas of skin depends upon the relative density of their networks of innervation and therefore upon the relative richness of the impulse patterns they can furnish. The finger tips have many more nerve endings per square centimetre than have the buttocks and the complexity of the nerve-fibre network varies at different levels below the surface of the skin as well as along it. The intestine has far fewer nerve endings than any skin area and so its activities only reach consciousness when they are very intense and widespread, for example, when it is obstructed. Cutting, pinching or burning small areas of intestine causes no sensation at all.

The localisation and quality of sensation depend upon the frequency, patterns and pathways of the nerve impulses resulting from the stimulation. There is continual monitoring, reorganisation and encoding of signal activity at all levels up to the cerebral cortex. This feedback is responsible for much of the variety of our response to stimulation. The emotionally disturbed often seem unusually sensitive to potentially painful stimuli while battle casualties may be pain-free after the first impact.

The neurone network

There are billions of neurones but only two or three millions of them are directly linked to the outgoing or efferent motor nerve fibres which end in the muscles. Of the rest some are sensory fibres leading directly from the outside world but the bulk of the neurones form the intermediate nerve network. There are 3000–5000 of these neurones to every motor neurone and they are responsible for all the subtleties of human response and activity. Seventy per cent of all the neurones in the central neuraxis are situated in the cerebral cortex and they are concerned with the reception, decoding, rerouting, storage, appreciation, programming and organisation of all the neuronal activity underlying all human thought, emotion and behaviour.

The neurones consume large amounts of energy in the maintenance of ionic concentration gradients and in the synthesis of structural proteins and of up to thirty different transmitter substances. Although the brain only forms 2 per cent of the total body weight it uses about 50 cm^3 of oxygen per minute, which is about one-fifth of the total bodily consumption. This utilisation rate is virtually constant throughout the twenty four hours. Most of the metabolic activity takes place in the cell body and there is a constant movement of material down the axon to the synapses. There is a fast system at a rate of 10–20 cm per day, which conveys the enzymes needed for transmitter synthesis, and a slow system (about 1 mm per day) for materials concerned with maintenance and repair of the axon itself.

Normally a neurone releases the same transmitter at all its synapses although there are exceptions. The various transmitters activate the different ionic transport channels in the postsynaptic cell membranes. Na^+ channels tend to reduce the membrane potential and are excitatory while Cl^- channels increase it and are inhibitory. The release of transmitters from the synaptic vesicles is triggered by the sudden inflow of Ca^{++} ions through their special channels. The interplay of synaptic activation determines the moment of firing of a particular neurone. After repeated activity synaptic transmission varies because of a fall in the number of transmitter molecules released. This underlies the phenomena of habituation, training and probably memory, depending upon the nature of the synapse – excitatory or inhibitory.

Reflex responses

The simplest response to a threatening stimulus is a single, involuntary, protective movement. Sensory signals travel along the nerve fibres from the skin to the spinal cord, where they excite a response which travels in motor neurones to the muscles and causes contraction to remove the part from danger. This type of response is a simple reflex and the pathway, skin → sensory neurone → motor neurone → muscle is a reflex arc. The junction, or synapse, between the *receptor* and the *effector* cell is situated in the reflex centre, usually in the spinal cord or brain stem. The *reflex arcs* for the simple reflexes of the body are pathways built into the nervous system before birth, ready to be developed by use. The stimuli concerned may originate inside the body, on its surface or at a distance. In many cases the sensory signals which evoke simple reflex responses also travel along other pathways to the brain, where they give rise to a variety of sensations, including pain.

A nervous system which could produce only simple reflex responses would be quite unable to activate the elaborate skills and abilities of the adult human body. The basic simple reflexes can easily be modified so that a variety of stimuli can evoke them. In dogs, if a whistle is blown at every meal time for some days, the whistle alone will then produce a copious flow of saliva. This is a *conditioned reflex* which will persist indefinitely if food is given occasionally to reinforce it. Conditioning can be very elaborate and, with suitable training, complex patterns of stimuli can be linked to reflex responses. Animals can distinguish very accurately between stimuli. A trained dog will respond only to a particular sound and will ignore any others. Conditioning is an important element in all animal, including human, learning and behaviour.

The movements and balance of the body are also based upon automatic, reflex adjustments of the muscles – *postural reflexes*. The sensory signals arise from the skin, joints, the muscles themselves, the eyes and the balancing organs in the ears. There are three main kinds of postural reflex – *static*, *righting* and *coordinating*. Signals arising from the skin, muscles and joints evoke static reflex contraction of the muscles of the legs, trunk and neck of the precise amount needed to maintain an established posture, such as standing. Any disturbance of this posture elicits fresh sensory signals which evoke righting reflexes tending to preserve the balance of the body.

Coordinating reflex activity is the integrated automatic muscular contraction that is essential for the effective performance of all normal activities such as walking, running and writing. Additional skills such as cycling or skating require the formation of chains of new reflex responses to movement and posture which then become permanent. The smooth, effective and controlled movement and progression of the healthy adult depends upon the continual succession of reflex responses of all types.

Analysis of simple reflex activity reveals certain basic features which help to explain the behaviour and responses of the intact animal. The responses are graded so that the speed of onset and intensity of a reflex response is directly related to the intensity of the stimulus. A minimal effective stimulus results in a minimal, localised response but as the stimulus is increased in intensity the response appears more rapidly, becomes more vigorous and may spread to muscles further away in the body. Stimuli too weak to evoke a reflex response may do so if they are repeated sufficiently quickly (*temporal summation*). The first stimulus momentarily prepares or 'sets' the reflex centre; subsequent stimuli add to this effect until the centre responds. Weak stimuli simultaneously affecting adjacent areas of the body may also add together to produce a reflex response (*spatial summation*).

The last neurone in a reflex arc is called the *final common path* (FCP). The FCPs for any given reflex arc are fixed and stereotyped but some of them may be used to form part of the FCP of other reflex arcs. The patterns of response are far less than the possible patterns of reflex stimuli. As the body is constantly receiving a great variety of stimuli it often happens that two reflex responses may be competing for the same final common path at the same time. In this event one stimulus will usually prevail over the other. The response which appears is said to be prepotent and it is always the one which has the greatest survival value for the organism. For example, if a painful stimulus is applied to the leg of a scratching dog the scratch reflex is replaced by the withdrawal reflex and the scratch reflex is said to be inhibited.

The nerve pathways on which reflex activity is based are laid down with a beautiful economy and functional efficiency so that the muscles needed to perform a particular action, such as withdrawal from a pin, contract in order and to the correct degree. At the same time the other muscles in the limb which could oppose the activity of contracting muscles are relaxed. All reflex responses are coordinated with each other and all muscular activity is a shifting mosaic of contraction and relaxation, exquisitely controlled. This is called the *principle of reciprocal innervation* and it ensures precise, accurate and effective reflex response to stimulation.

Reflex responses are easily fatigued, the reflex time becomes longer and the response smaller in amplitude. Such fatigue must be due to changes in the reflex centre because the peripheral nerve fibres can still respond to stimuli normally long after reflex activity has failed. Central fatigue can result from mere repetition of the reflex response and is presumably essential if chains of reflex responses, once set into activity, are not to persist indefinitely. Alterations of blood, oxygen or glucose supply, changes in the pH of the blood (pp. 17, 98) and drugs such as alcohol, caffeine and anaesthetics all modify reflex activity and change reaction time by their actions on the nerve cells and their synapses.

Voluntary movement

Voluntary, or purposive, movement does not occur in isolation but in response to the needs of the individual. Each movement is related to previous events and is modified by estimates of future developments. The nervous system operates as a coordinated whole, and it is constantly re-arranging and altering the overall activity of the body so as to produce the most effective adaptation to changing circumstances.

Signals are continually flashing around the whole of the nervous system and about three million are generated every waking second. At the nerve endings these signals produce a transient change in the processes or cell bodies of other neurones. Nerve impulses in suitable patterns of space and timing will fire a fresh impulse in the target neurone. Elaborate signal patterns surge constantly throughout the brain and spinal cord. The incessant bombardment of the surfaces and processes of the neurones continually change their state of excitability. Much of this traffic is insufficient to produce a new signal, but it can still increase or decrease the readiness of the cell to respond to the next signal cluster that reaches it.

In addition to these transient, fluctuating changes in excitability there are presumably more enduring 'preferred pathways' associated with the development of new or more complicated reflex arcs, with learning and the phenomenon of memory. Short-term memory is easily disturbed by head injury, hypothermia and anaesthesia. It is probably basically an electrical phenomenon related to excitability.

Long-term memory may be due to the formation of new synaptic endings producing new neuronal circuits or to the enlargement of existing synaptic endings. The neurone is thus 'set' or prepared to react preferentially to a familiar group of signals.

A voluntary movement can be thought of as starting in the motor nerve cells of one cerebral hemisphere in response to a unique bombardment of nerve signals streaming in from many sources. The pattern and rhythm of the cells firing will be related to their previous activity and to the extent and type of movement to be performed. The trains of signals set up pass to the motor nerve cells in the spinal cord, which are themselves modified by their previous patterns of activity and by the constant arrival of signals from all parts of the nervous system. These signals arrive in sensory nerve fibres entering the spinal cord at the same level, in extrapyramidal nerve fibres from the basal ganglia and cerebellum and in connecting fibres from nearly all parts of the spinal cord. Some of these signals promote activity – excitation – in the receiver nerve cells, but others tend to prevent or inhibit activity. The balance between excitation and inhibition of the motor nerve cells of the spinal cord, as anywhere else in the nervous system, determines the pattern of response that is finally produced. In the case of voluntary movement the signal patterns have to be precisely adjusted to produce the exact rate and degree of muscular contraction required. This must be smoothly and accurately grafted on to the background of automatically adjusted postural reflex activity of the muscles. The power exerted by motor units firing at different rates, sequences and sites provides the essential basis of all effective human activity. The overall, supervisory control of total motor performance depends upon the coordinated activity of the cerebellum and basal ganglia.

The coordination and control of movement

Precise, effective movement requires the graded contraction of various muscles in succession. Although we speak of 'voluntary' movement, we are not consciously aware of contracting individual muscles. Yet the smooth, accurate and forceful activities of the body involve continual readjustment of the postural reflex activity of the muscles. The balance and set of the trunk and limbs are constantly modified to provide the most efficient background for present and intended movements. This depends upon the passage of a stream of accurate information into the nervous system which enables the necessary corrections and adjustments to be made.

The movement and position of the joints, the length and rate of stretch of the muscles, the load on the tendons and the position and movement of the body in space are all monitored by special nerve endings. They give rise to a flood of nerve impulses, which continually pass to the brain and spinal cord, modifying the patterns of motor activity signalled out to the muscles. Information from the muscles and tendons passes mainly to the cerebellum, which also receives signals from the cerebrum in relation to all the motor patterns sent down to the spinal cord and muscles. The cerebellum can thus compare the performance of the muscles with the patterns initiated by the cerebrum. Any error or difference between them is detected by the cerebellum, which sends further impulses up to the cerebrum, causing corrective impulse patterns to be sent to the muscles. These may increase or decrease the degree of activity of the different muscles. The cerebellum assesses the rate of change of the parts of the body and automatically predicts and directs the degree and rate of contraction of the muscles required to produce the intended action. The first attempts at unfamiliar movements are always clumsy and ill-adapted, but the development of skill is accompanied by the building up of new patterns of postural reflex function and of cerebellar control. Skilled, learned activity improves with repetition, but the processes involved are entirely subconscious. They can be easily disturbed by emotion but are little affected by efforts of 'will'. In fact, conscious attempts to analyse the muscular sequences of a skilled activity like a golf shot can disrupt the performance completely.

Speech and writing

Rapid, accurate communication by speech or writing is a unique human characteristic. The understanding of speech or writing and the power of expression in words are closely related activities that involve extremely elaborate patterns of nervous activity. They also require extensive, readily accessible memory stores. Speech and thought are closely linked functions of large regions of the brain and cannot be pinpointed as originating in discrete areas.

The degree of activity of the nerve cells of the brain is directly dependent upon the ceaseless inflow of sensory signals from all parts of the body. The resting individual in a quiet, darkened room quickly drifts off to sleep. The state of consciousness and the processes of thought are related to the pattern of intensity of the inflow of sensory information. The significant patterns of sound that comprise speech are set up by long, serial patterns of related muscular contractions organised and carried out by the nerve cells of the cerebrum. Words are symbols with a significance established by experience. Meaning is conveyed partly by the patterns of words but also by changes in the speed and rhythm of speech, by the tone of voice, by gesture and by bodily attitudes. In fact, the most striking feature is the subtlety and economy of speech and movement that the brain employs to communicate a wealth of information and emotion.

Sleep

Adults need six to eight hours of sleep every twenty-four hours, and if deprived of it they become increasingly inefficient and distressed. Yet the function of sleep is dimly understood. During sleep, bodily activities are modified and the metabolic rate falls. The pulse rate, blood pressure, body temperature, breathing rate and heart rate all fall. The reduced breathing retains some carbon dioxide in the body. This is counteracted by the kidneys, which secrete a more concentrated, acid urine during the night.

The mechanism of sleep is obscure but it is evident that a constant bombardment by sensory nerve impulses conduces to wakefulness. Even during sleep certain nerve pathways are still accessible. For example, the mother will awake at once if her child whimpers but may be undisturbed by traffic noise. Reduced, but not abolished, brain activity seems to be the essential condition of sleep.

The electrical activity of the brain (EEG) changes during sleep. At first the typical waking alpha rhythm of 8–12 cycles per second is replaced by slower delta waves interspersed with 'spindles' of faster beta waves. At intervals during the night these appearances are interrupted by spells of more normal rhythm at lower voltage. These spells last for 15–20 minutes and are accompanied by changes in blood pressure and breathing and by rapid eye movements (REM sleep). Dreaming and nightmares occur during REM sleep. It is thought to be due to activity in the brain stem reticular formation (p. 142).

13

The Autonomic Nervous System

The functions of the internal organs of the body, which are normally automatic and unconscious, are largely controlled by the autonomic nervous system. This is an integral part of the central nervous system which is usually considered separately on anatomical grounds. It controls the function of various glands, the movements of the heart and gastro-intestinal tract, the activity of the smooth muscle of the blood vessels, bronchi and urinary bladder, the maintenance of blood pressure and the state of the skin. It ensures the smooth running of all these complicated, hidden processes and although its activity is entirely below the level of consciousness and involuntary, it is at the same time closely related to the emotional state of the individual. The extensive physical reactions to embarrassment, fear and anger are all mediated by the autonomic nervous system.

It is divided into two parts which are constructed and function in quite different ways (Fig. 13.1). One part – the *sympathetic* nervous system – produces a series of changes which prepare the body for vigorous activity like fighting or running away. Its action is augmented and prolonged by adrenaline, which is secreted by the medulla of the suprarenal gland whenever the sympathetic nervous system is working briskly. The result is a faster heart beat and an increase in output from the heart. Blood vessels supplying less vital organs like the skin are constricted so that the maximal supply of blood reaches the vital organs such as the brain and active muscles. The movements of the intestine and digestion are diminished or stopped, but liver glycogen is broken down as a source of energy. At

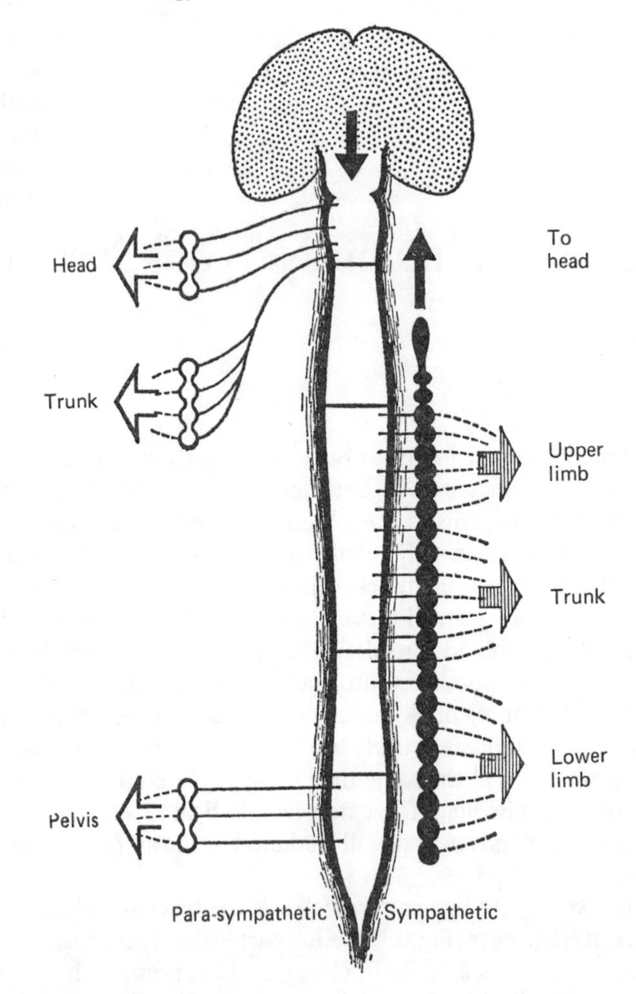

Fig. 13.1 The overall arrangement of the autonomic nervous system

the same time, the metabolism of every cell in the body is increased and more heat is produced. In addition to such widespread general effects, selective actions can also be produced by the sympathetic nervous system. It constricts the blood vessels in the lower half of the body when a person stands up. Pooling of blood in the legs and consequent dizziness or even fainting is prevented.

The other part – the *parasympathetic* nervous system – works towards the conservation of the body and the restoration of its energy. In general, its actions oppose those of the sympathetic division and the two work harmoniously together to regulate the internal workings of the body. It slows the heart, diminishes the circulation and increases intestinal function. Ninety per cent of the parasympathetic nerve fibres are contained in the tenth cranial nerve – the *vagus*, which supplies all the organs of the chest and abdomen.

Organ	*Effect of sympathetic activity*	*Effect of parasympathetic activity*
Eye pupil	Dilated	Constricted
Blood vessels:		
1 Abdominal	Constricted	Nil
2 Muscle	Dilated	Nil
3 Heart	Dilated	Constricted
4 Skin	Constricted	Dilated
Bronchi	Dilated	Constricted
Intestinal Wall	Decreased activity	Increased activity
Bladder Wall	? Inhibited	Contracts
Sweat Glands	Increased	Nil
Blood Sugar	Increased	Nil
Metabolism	Increased	Nil

Table 13.1 The effects of autonomic nervous activity

Autonomic reflexes

The functional unit of the autonomic nervous system is exactly the same reflex arc as was described in Chapter 12 for the rest of the nervous system. These reflex pathways are laid down during foetal development and the responses are therefore constant and stereotyped. Autonomic reflexes are responsible for local vasomotor responses to changes in temperature, for reflex emptying of the bladder and rectum and for the changes in sweating and in blood pressure which occur in response to stimuli arising in the intestines and other organs.

The essential anatomical feature which characterises the autonomic nervous system is the situation of the junction between the synaptic ending of the connector nerve fibre and the cell body of the effector neurone. In the rest of the nervous system all the synaptic junctions lie within the central neuraxis but in the autonomic nervous system the final junction lies outside the central neuraxis in a series of special aggregations called *ganglia*.

The details of the central control of the autonomic nervous system are not fully understood. It is to some extent influenced by the cerebral cortex and this may explain the physical accompaniments of emotional states, like blushing, pallor, sweating, increased pulse rate and rising blood pressure. The autonomic nervous system is more directly regulated by the complicated nerve cell masses, or *nuclei*, situated in the medulla, brain stem and hypothalamus (p. 143). The hypothalamic nuclei are particularly concerned with the regulation of the metabolism, blood sugar level, water balance and body heat. The nuclei in the medulla and brain stem are rather more widely scattered but they regulate the behaviour of the respiratory and cardiovascular systems and are sometimes referred to as the respiratory and vasomotor centres.

The sympathetic nervous system

The cell bodies of the connector neurones of the sympathetic nervous system lie in the grey matter of the spinal cord from the levels of the eighth cervical to the second lumbar segments. The nerve fibres from these cells leave the central nervous system in the anterior nerve roots, but they split off from them to end in 22 paired sympathetic ganglia, which are linked together to form the two sympathetic nerve chains running down the back wall of the body cavity. These primary efferent nerve fibres are called preganglionic fibres and they end in close approximation to numerous secondary nerve cell bodies in the sympathetic ganglia. The secondary cell bodies give rise to the secondary efferent or postganglionic nerve fibres (Fig. 13.1).

The postganglionic nerve fibres leave the sympathetic nerve ganglia and return to the spinal nerves with which they are distributed to their various destinations in the sweat glands, the muscles around the hairs of the skin, the blood vessels and smooth muscle of the viscera. The sympathetic nerve supply to the head arises from the upper end of the sympathetic nervous system and travels in the walls of the arteries going to the head (Fig. 13.1).

The parasympathetic nervous system

The autonomic nerve fibres which originate from cell bodies lying in the brain stem and in the sacral part of the spinal cord form the parasympathetic nervous system. There are pre- and postganglionic nerve fibres in the same way as in the sympathetic nervous system, but the ganglia are situated much more distally, in the organs they supply. Preganglionic nerve fibres arising from cells in the mid-brain travel out with the third cranial nerve fibres to the ciliary ganglion, which gives rise to postganglionic fibres supplying the pupil (iris and ciliary muscles). Other preganglionic fibres arise from nerve cells in the medulla and travel with the nerve fibres of the seventh and ninth cranial nerves to innervate the salivary and other glands of the head and neck.

Other cell bodies in the medulla give rise to the largest autonomic nerve in the body, the vagus nerve (tenth cranial nerve), which contains all the preganglionic fibres to the many ganglia lying in or near the organs of the chest and upper abdomen. Short postganglionic nerve fibres run from these ganglia to supply the heart, intestines, pancreas and spleen. Ninety per cent of all the parasympathetic nerve fibres in the body travel in the vagus nerve.

The remainder of the parasympathetic nervous system consists of preganglionic fibres arising in the sacral region of the spinal cord and eventually terminating in pelvic ganglia, from which postganglionic fibres travel to the smooth muscle of the bladder, rectum and genitalia. These fibres form the outgoing or efferent side of the various reflexes concerned in micturition, defaecation and sexual activity.

Autonomic transmitter substances

The passage of the nerve impulses between autonomic neurones and from the final neurones to the target organs is achieved by releasing chemicals called transmitter substances. The transmitter substance released at all junction points in the autonomic nervous

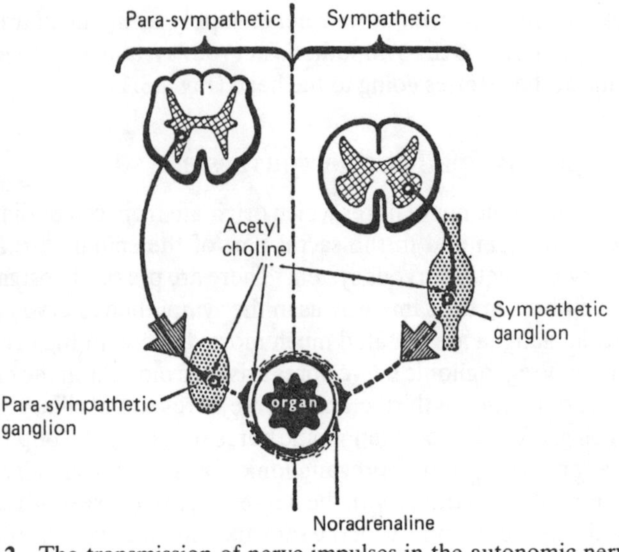

Fig. 13.2 The transmission of nerve impulses in the autonomic nervous system

system, except the final endings of the sympathetic division, is *acetyl choline* (Fig. 13.2). Each jet of acetyl choline liberated at a nerve ending can produce only one response in the next cell because it is rapidly destroyed by an enzyme, cholinesterase. Similarly, the transmitter at sympathetic nerve endings is noradrenaline, which is quickly destroyed by the enzyme amine oxidase. After sympathetic nerve activity, significant amounts of noradrenaline and adrenaline may enter the blood stream and affect distant adrenergic nerve endings. Similarly, noradrenaline and adrenaline liberated by the suprarenal medulla (p. 183) can influence sympathetic nerve endings via the blood stream. For this reason the sympathetic nervous system and the suprarenal medulla are usually regarded as a functional unit.

Autonomic function

Normally both sections of the autonomic nervous system are working at a low level of activity, sending a steady, basal, stream of impulses to the sympathetic and parasympathetic nerve endings. This is sometimes called the *tone* of the autonomic nervous system and it is responsible for the maintenance of normal levels of activity in the various structures it supplies. Deviations from the normal are countered by increase or decrease in the activity of the appropriate part of the autonomic system. Essentially, the sympathetic and parasympathetic nervous 'systems' are merely the final reflex pathways by which the nervous system modifies and controls the basal functions of the body, such as respiration, circulation and metabolism. The sympathetic component ensures the 'homeostasis' (p. 2) of the internal environment of the body, while the parasympathetic nervous system protects and restores the physical resources of the organism.

14

Special Senses

Vision

Ordinary white light is a mixture of red, orange, yellow, green, blue and violet light. The colour of objects depends on the mixture of coloured light that they reflect. Black objects absorb all the light reaching them and white ones reflect it. Light entering the eye stimulates sensitive nerve cells in the retina and gives rise to nerve signals which pass along the optic nerves and, after relaying, reach the back of the brain. The pattern in time and space of these signals is the basis from which our awareness of shape, size and colour arises. From these patterns and from the minute adjustments of the eyes themselves we make rapid judgements of movement, distance and relationship of objects.

Visible light consists of electromagnetic energy travelling at very high speed in a straight line (almost 300 000 km/s). For our present purpose it can be regarded as being composed of a series of waves. The very minute distance between the crests of successive waves is known as the *wavelength*. Each colour we can recognise has a specific wave length, but normally light reaching the eye consists of a mixture of different wavelengths. Light is bent when it passes from one substance to another, e.g., from air to glass, and the degree of bending is proportional to the wavelength. If white light is passed through a triangular piece of glass (a prism) it emerges as a coloured band, or spectrum, comparable to a rainbow, because the lights of different wavelengths are bent to different degrees. The colours of the spectrum are in order red, orange, yellow, green, blue and violet.

An object appears to be white if, on striking its surface, light of all wavelengths is reflected to the eye of the observer. Objects are seen as coloured when they reflect light of the wavelengths corresponding to that colour or mixture of colours but absorb the rest. Any of the hues which we can, with perfect vision, distinguish, can be produced by suitable blending of light of different wavelengths. The appreciation of colour is also modified by the intensity of the light and by contrast effects when colours are placed side by side. Black looks blacker against a light background and white seems whiter if surrounded by a dark area. This phenomenon of simultaneous contrast is of great importance when planning colour schemes because each colour must be considered in relation to its background.

Structure of the eye

The eye is roughly spherical with a transparent window, the *cornea*, in front and a large nerve tract, the *optic nerve*, behind (Fig. 14.1). It is divided into two unequal parts, by the *lens*, which has a diaphragm containing pigmented muscle, called the *iris*, in front of it. The iris has a central circular hole, the *pupil*, which varies in size, and this

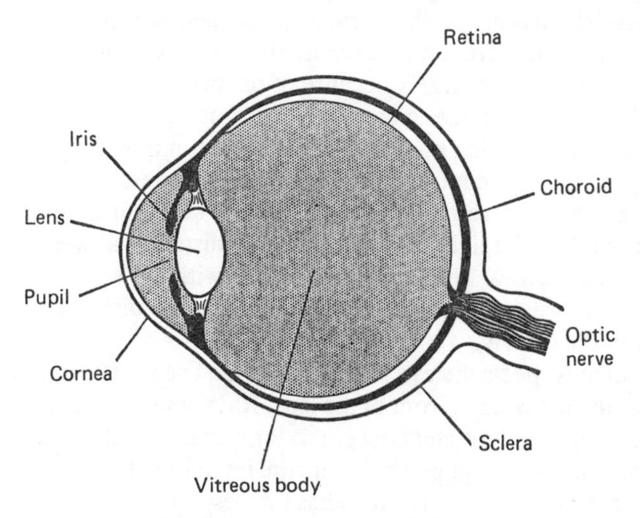

Fig. 14.1 Cross-section of the eye

controls the amount of light passing through it to the *retina*. In bright light the iris muscle contracts and the pupil is small; in semi-darkness the pupil is fully dilated.

The space between the lens and cornea contains a watery fluid, the *aqueous humour*, which is rather like blood plasma without the protein. This space is divided into anterior and posterior chambers by the iris and the pressure of the aqueous humour in them is normally maintained at 3 kPa (22.5 mmHg). The aqueous humour is formed by filtration from the capillaries of the ciliary body in the posterior chamber in a similar manner to the formation of the glomerular filtrate in the kidney (p. 95). It is renewed every four hours and is absorbed from the anterior chamber at a rate comparable to its formation so that the pressure inside the eyeball is maintained and the eye keeps its shape. The larger space behind the lens is filled with a soft jelly called the *vitreous humour*.

The back of the eye is lined with a pigmented coat inside which is an elaborate arrangement of light-sensitive structures, the *rods* and *cones*, and a rich network of nerve cells and fibres (Fig. 14.2). This forms the retina, which responds to the different light energies and transforms them into nerve impulses which pass information back to the brain where the experience of vision occurs.

Light is brought to focus on the retina by the combined bending powers (refraction) of the cornea, the aqueous humour and the lens. In man light from objects more than 60 metres away forms an image on the retina largely because of the bending which occurs at the air–corneal surface. Nearer objects are kept in focus by variation in the refracting powers of the lens (accommodation). This is achieved by the contraction of the ciliary muscles which are attached to it and can change the curvature of its surface. The eye can thus adjust for light coming from great distances as well as from a few inches away. The lens grows harder with age so that the eye loses its power of accommodation and reading difficulties often occur in the elderly.

In some respects the eye is like a camera. The shutter is comparable to the iris lying in front of the lens. Rays of light from an object pass through the lens and are bent to form an inverted image on the retina. Variation in pupil size alters the input of light, concentrates it at the centre of the lens which has the best optical properties and improves the depth of field by reducing the aperture.

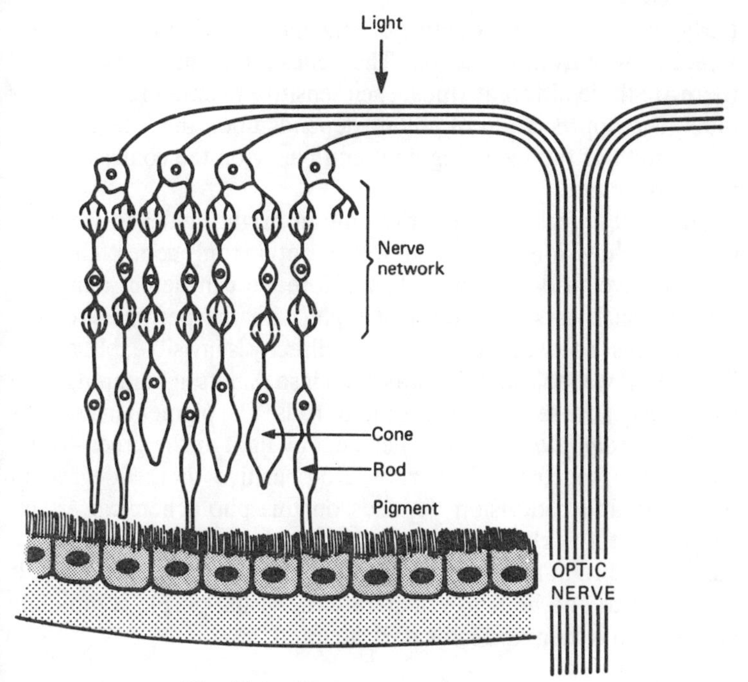

Fig. 14.2 The structure of the retina

STRUCTURE OF THE RETINA

The back of the eye is lined by the retina, which consists of three main layers (Fig. 14.2).

1 The innermost layer which the light first penetrates is a network of nerve cells with fibres which finally run together and leave the eye to form the optic nerves.
2 Behind the nerve network there is a layer of light-sensitive cells of two different types, the rods and the cones.
3 The hindmost layer of cells forming a backing to the rods and cones, containing a black pigment called *melanin*. This absorbs any excess light and prevents reflection across the eye.

There are about seven million cones and one hundred and twenty million rods in each eye. Most of the cones are massed together at one tiny spot, the *fovea*, at the back of the eye, which is the central focusing point of the whole optical system. The rods are spread fairly uniformly over the retina but the number of cones diminishes steadily away from the fovea. The cones are colour sensitive and there are three different kinds, each sensitive to either red, green or blue. The appreciation of colour depends upon the intensity and wavelength of the incoming light and partly on the contrast in the patterns formed.

The cones work satisfactorily only in bright light and we have poor twilight vision because our fovea contains only cones. The rods are sensitive to dim light and therefore we can detect dim light sources, such as stars, better at night with the periphery of the retina. This is why the star we look at directly is invisible, but comes into view if we look slightly away from it so that its light impinges on some part of the retina which contains rods. The pigment in the rods is called *visual purple* and is bleached by light at the green–violet end of the spectrum. The rods are sensitive to the degree of bleaching and rod vision depends on this photochemical effect. Bright light quickly bleaches all the visual purple, so that we are dazzled by bright lights on a dark night and cannot see properly. It takes half to one hour to regenerate the visual purple completely and the eye is then said to be dark adapted.

Colour blindness

Difficulty in differentiating colours is fairly common and may affect up to 20 per cent of the population to some extent. The most frequent anomaly is confusion between red and green. This is inherited as an X-linked recessive trait (p. 27) and affects 8–10 per cent of males. It also occurs in homozygous females. The deficiency is due to the basic reduction in sensitivity of one or more colour response in the cones.

Visual pathways and visual fields

The complicated retinal nerve network sorts out, sifts and organises the responses of the individual rods and cones and transmits patterns of nerve impulses along the optic nerves to the brain. There are more than one million nerve fibres in each optic nerve and it is believed that each cone is connected to an individual nerve fibre but as many as three to four hundred rods may form a functional unit sharing a single optic nerve fibre.

The optic nerves run back into the skull and join together in front of the pituitary gland to form the *optic chiasma*, in which some of the nerve fibres cross to the other side. Nerve impulses which arise from the nasal half of each retina cross over in the optic chiasma and finally reach the occipital cortex on the opposite side of the brain. Impulses arising from the temporal halves of the two retinae do not cross over and therefore end in the brain on the same side (Fig. 14.3).

The camera-like effect of the lens of the eye means that objects to one side of the body will form images on the retina on the opposite side. Objects to the left front produce images on the right half of each retina and impulses from these areas reach the right occipital cortex (dotted lines). In this way objects appearing on one side of the body are reported to the opposite side of the brain. This means that the right occipital cortex is responsible for 'seeing' objects to the left-hand side of the body and the left occipital cortex 'looks' to the right.

Eye movements and gaze

The eyes are embedded in the bony orbits, protected and supported by cushions of fat and connective tissue. Six long, thin ribbons of muscle run from the orbits to the outer surface of each eye. They contract together so that the eyes move in unison with great precision and speed. Their movements are so well correlated that the images on the two retinas always correspond with each other. The eye muscles are controlled by motor-nerve fibres, arising from nerve cell groups (nuclei) lying in the brain stem, which form the third, fourth and sixth pairs of cranial nerves. The ocular motor nuclei are connected by networks of nerve cells which are concerned with coordinating their activity and organising their patterns of discharge. Although eye movements are partly under voluntary control, most eye activity is an automatic response to a variety of stimuli such as vision, smell, hearing, touch and even memory.

The detection of movement is essential for survival in the wild and the eye has evolved primarily as a detector of movement. The periphery of the retina only responds to movement which reflexly turns the eye towards the object. When searching the eyes move in tiny, quick jerks but when following an object they move slowly. This ensures that the stimulus to the retinal receptors is ever changing. A constant stimulus soon loses its effect through fatigue

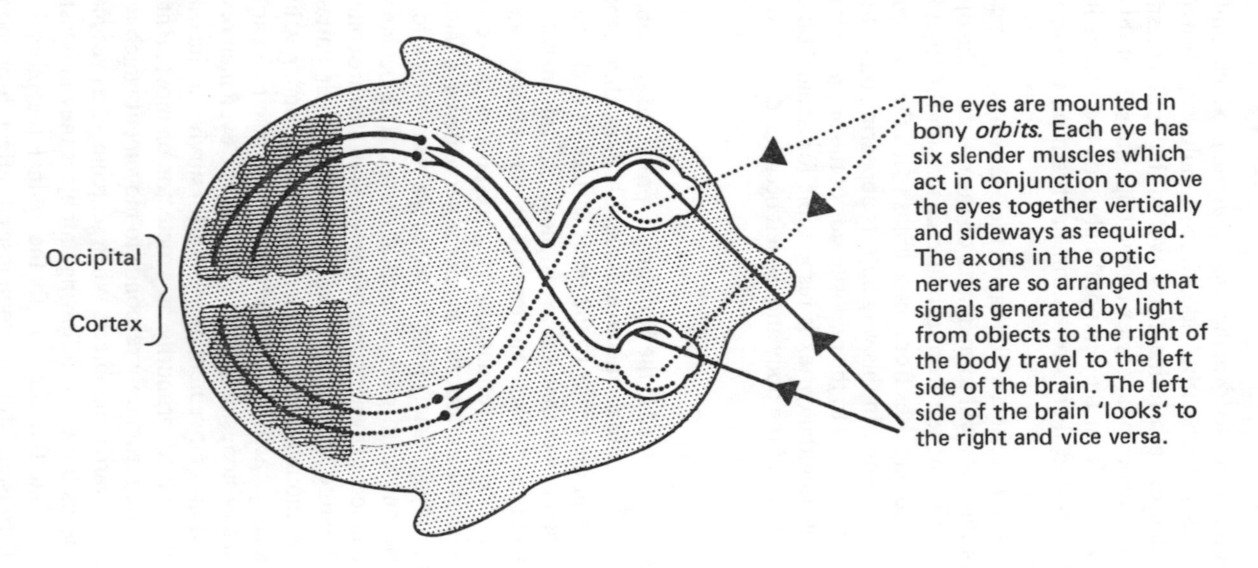

Fig. 14.3 Visual pathways in the brain

or adaptation. The receptors respond to moving boundaries between objects and to changes in the distribution and intensity of light.

The responses to stimuli are coded into patterns of neural activity in the retinal nerve network. These patterns travel via the optic nerve to the occipital cortex where, in an entirely mysterious manner, they give rise to the sensation of vision. From here further neural patterns pass on to the brain stem, where they are correlated with information from the muscle spindles and other receptors in the head, neck, eye muscles and labyrinths. There is then a constant outflow of nerve impulses which precisely controls eye movements and position ensuring a completely stable and significant visual image.

Hearing

Sound is caused by waves of compression travelling through the air or other material which is in contact with a vibrating or moving object. The *intensity* of sound depends upon the amplitude of the waves, the *pitch* depends upon their frequency and the *quality* depends upon the combination and relationships of the waves. The ear converts sound energy into nerve impulses, and in humans vibrations ranging from 30 to 20 000 per second can be detected. Low frequencies produce low notes and high frequencies the high notes. A musical note consists of a basic frequency plus a number of overtones (harmonic frequencies) which are related mathematically to the basic frequency.

Sound waves are channelled by the *external ear* along the *auditory canal*, which ends at a tightly stretched membrane, the *ear drum* (Fig. 14.4). The *middle ear* beyond the ear drum is connected to the pharynx by the *Eustachian tube*, which enables the pressure on the two sides of the drum to be kept equal. Vibrations set up in the ear drum by sound waves are transmitted by a system of linked bony levers or ossicles to the *inner ear* or *cochlea*. This is a coiled tube about 2.5 cm long filled with fluid, or perilymph, similar to cerebrospinal fluid. It is divided lengthways by a strong *basilar membrane* upon which lies the sensitive *organ of Corti*. The two sides communicate at the inner end of the membrane so that the resting pressure is equal overall.

The ear drum and ossicles concentrate the energy of the sound

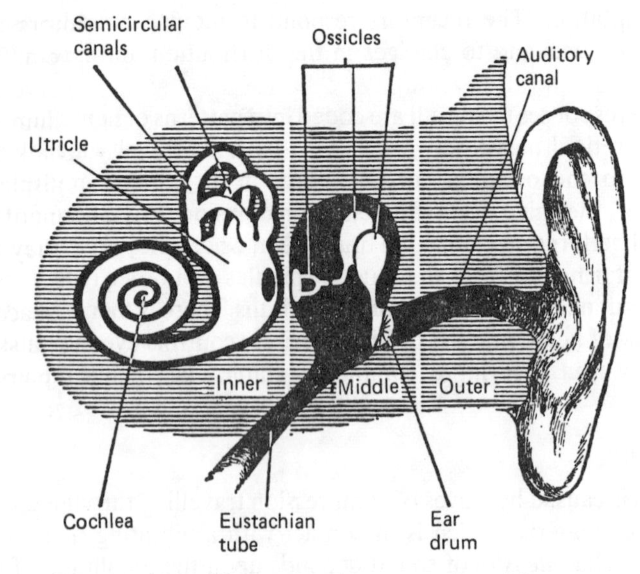

Fig. 14.4 Diagrammatic cross-section of the ear

waves on to the cochlea and the vibrations pass through its fluid to the basilar membrane. It has different resonating properties along its length, responding to high frequencies at the middle-ear end and to low frequencies at the other. The actual receptors are hair-like structures in the organ of Corti which respond to movement of the basilar membrane by generating nerve impulse patterns which travel along the auditory nerve to the temporal lobes of the brain. The nerve-impulse patterns are interpreted in the cortex of the temporal lobe and related to the overall pattern of sensory activity in the brain.

Organs of balance

The organs of balance are an important part of the mechanisms that preserve the balance of the body. Each ear contains a *vestibular apparatus*, which detects the position of the head in space and also signals any change in its position or movement. Each one consists of a chamber called the *utricle*, which communicates with the cochlea, and three *semicircular canals*. The whole system is full of thick fluid, endolymph, and each part is supplied with sensory receptors consisting of stiff hairs with nerve fibres entwined among them. The nerve fibres form the vestibular portion of the auditory nerve and run to the centres for balance in the brain stem and cerebellum.

The sensory receptors in the utricle have tiny lumps of chalk, the *otoliths*, entangled in their hairs. The weight of the otoliths displaces the hairs, depending upon the position and rate of movement of the head. They respond sharply to sudden changes, and if they are overstimulated motion sickness may result.

The semicircular canals are arranged in three planes of space at right angles so that they complement each other. When the skull moves suddenly in any direction, the fluid in one or more pairs of canals will lag momentarily. The inertia of the fluid bends the hair-like receptors in proportion to the degree and amplitude of the movement so that patterns of nerve impulses are transmitted to the brain. The nerve fibres from the nerve networks join to form the vestibular portion of the auditory nerve, which runs to the vestibular nucleus in the brain stem. Fresh neurones run from the vestibular nuclei to the cerebellum, to provide information for the coordination of movement and balance, and others go directly to the ocular motor nuclei and to the spinal cord to give reflex control of eye movement and maintain the posture of the head and neck.

The virtual absence of gravity during space travel produces vestibular disturbances. Motion sickness and disorientation are frequent problems.

Taste

The mucous membrane of the tongue contains specialised receptors – *taste buds* – stimulation of which can evoke four different tastes, bitter, sweet, salt and sour. The nerve fibres from these receptors convey the nerve impulses by circuitous routes to the lower end of the sensory area (p. 145) in the parietal portion of the cerebral cortex. This fairly crude mechanism is augmented by multiple patterns of nerve impulses in other pathways, especially vision and smell, to produce the refinement of taste.

The flavour of food and drink arises from a combination of taste, smell, texture and temperature. Fatigue due to adaptation develops rapidly.

Smell

Our sense of smell is much more sensitive than our sense of taste and we can detect some chemicals at a dilution of one part of ten million parts of air. The receptors are nerve cells situated in a small area of mucous membrane inserted in the base of the skull high in the nose. Nerve fibres from these receptors enter the skull and run back in the *olfactory tract* to terminate at the inner side of the temporal lobe of the brain. The sense of smell is very sensitive, but it tires (adapts) very readily, although only for the particular odour present. This helps workers in offensive atmospheres that nauseate the uninitiated. Taste and smell are of greater importance in animals than in man because of the necessity for rapid identification of food, danger or other animals of the same species.

The skin

The elaborate network of sensory nerve endings in the skin make it one of the most sensitive sense organs in the body. Structurally there are many different sensory receptors which are thought to have different functions; that is, to be particularly sensitive to different forms of energy such as heat, cold, touch, pressure and pain-provoking stimuli. The acuity of sensation and the accuracy of its localisation are dependent upon the density of innervation of the skin. Sensitive areas like the finger tips have a much larger number of nerve fibres and endings per square millimetre than relatively insensitive areas like the buttocks.

The skin forms about one-sixth of the total body weight and has an area of 1.6–1.8 square metres in an average person. It consists of an outer layer of cells, the *epidermis*, resting on a framework of muscles, nerves, blood vessels and connective tissue called the *dermis*. It provides a water-resistant, protective covering for the deeper tissues but it soon loses this quality if it is neglected. It contains *sweat glands*, which secrete water and other substances (p. 65), and *sebaceous glands*, which produce a fatty material called *sebum*. Sebum helps to waterproof the skin and keep it supple. Lack of cleanliness, wetness, abrasion, infection and pressure which interferes with its blood supply all rapidly lead to injury and even local death of the skin.

The skin has a rich blood supply which controls its temperature and, to a large degree, its colour. The arterioles reaching the skin surface each lead into about twenty capillary loops, which course out at right angles to the skin and lead back again into minute venules. The skin forms a reservoir of blood and after blood loss, or in shock, the arterioles are intensely constricted and the skin becomes pale and cold. When the body is chilled the arterioles constrict so that heat is conserved but in hot conditions they dilate and the skin becomes warm and red. In very hot weather the skin may feel hot and remain pale because blood is then shunted directly from the arterioles to the veins, leaving the superficial vessels empty. This mechanism allows large quantities of blood to reach the body surface and give off heat. Conversely, stagnation of blood in the skin, because of circulatory failure or obstruction, produces a cold, blue skin.

The colour of the skin also depends upon the presence of a pigment, *melanin*, in special cells in the deeper layers of the epidermis. The amount of melanin is related to the colour of skin and hair, but it is also increased by exposure to ultraviolet light rays in sunshine and exposed surfaces like the face and hands are normally darker than the rest of the body.

Precursors of vitamin D are stored in the skin. Under the influence of ultraviolet rays in sunlight they are mobilised and enter the blood stream. They are eventually transformed into metabolically active vitamin D in the liver and kidney (p. 62).

Hair and nails

Both hair and nails are modified forms of epidermis. They are of some importance in protection of the underlying tissues.

15

Endocrine (Ductless) Glands

Many activities are influenced, controlled and coordinated by chemical messengers – *hormones* – carried in the blood stream. Many of these hormones are produced by special organs called ductless glands because their hormones pass directly into the blood stream instead of long ducts. Some glands, like the *thyroid*, produce a single hormone that influences every cell in the body. Others, like the *parathyroid* glands, produce a hormone that is only concerned with a single metabolic activity – calcium and phosphorus metabolism. Glands like the *suprarenal* produce numerous hormones with multiple activities. The various glands often have considerable influence upon each other but the *pituitary* gland directs and coordinates the activities of the ductless glands as a whole (Fig. 15.1).

Some hormones are relatively simple chemical compounds, others are very complicated. They may be active at very low concentration, e.g., *adrenaline* from the suprarenal gland is effective at a strength of one part in four hundred million parts of water. The pituitary, thyroid, parathyroid and suprarenal glands, the stomach, intestines and sex organs all produce hormones. The processes and systems influenced by them include growth, metabolism, circulation of the blood, heat production, response to infection and other stress as well as the maintenance of the correct bodily levels of fluids, salts, sugar and proteins. The functions of the ductless glands are summarised in Table 15.1

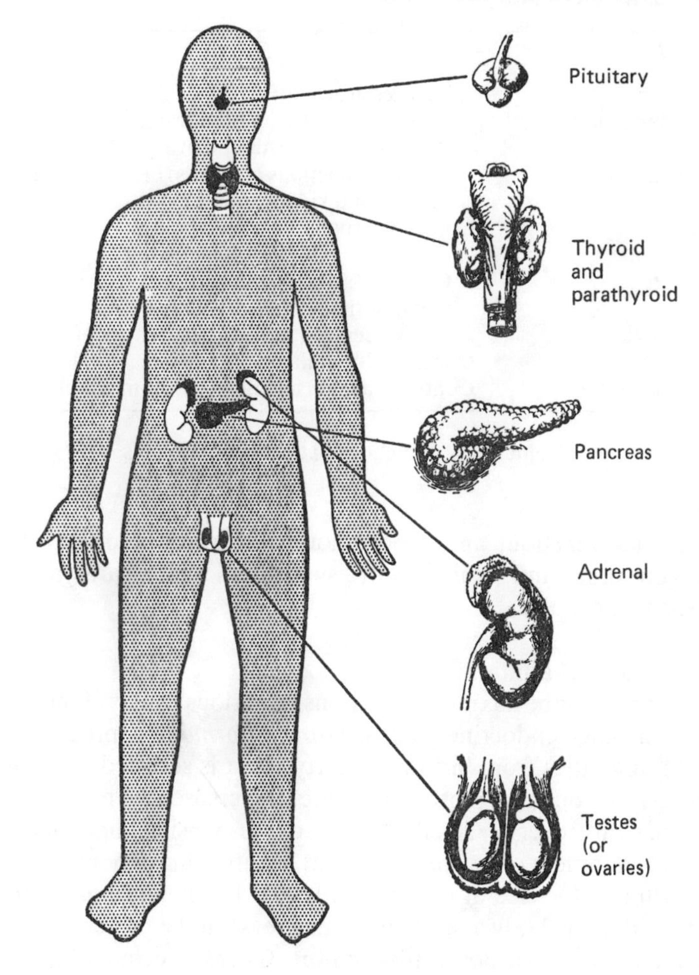

Fig. 15.1 The endocrine glands

The pituitary gland

The pituitary gland is a reddish-grey organ, situated in a cavity in the sphenoid bone at the base of the skull. It is attached to the brain by a slender stalk, weighs 0.5 g and measures 12 × 8 mm. It consists of an anterior lobe, which is derived from the roof of the mouth and has a

Gland	*Function*
Pituitary	Influences growth of all cells.
Anterior lobe	Controls the activity of suprarenal and thyroid glands, testis and ovary.
Posterior lobe	Controls functions of uterus, breast, blood vessels, and kidney tubules.
Pancreas	Controls carbohydrate metabolism.
Suprarenal	
Cortex	Controls, salt, protein and carbohydrate metabolism.
Medulla	Reinforces sympathetic nervous system.
Thyroid	Controls all cellular metabolism.
Parathyroid	Controls calcium and phosphorus metabolism.

Table 15.1 Functions of the ductless glands

glandular function, and a posterior lobe, which is made up of nervous tissue and receives both a sympathetic and a parasympathetic nerve supply.

Functions of the anterior lobe

The anterior lobe has multiple actions on various tissues of the body and on other endocrine glands. *Growth hormone* is produced in large quantities from birth to puberty but it is secreted in smaller amounts throughout most of adult life. It increases the number and size of nearly every cell in the body and is responsible for regulating the development of height and weight, lengthening of bone, muscle growth and the size of the organs. Under its influence nitrogen is retained in the body and there is an increase in the protein content of the muscles but not in their power. Growth hormone also has important metabolic effects which in general are opposite to those of insulin. For example, it causes a rise in the blood sugar level and relative insensitivity to the action of insulin.

The pituitary hormones which affect other glands are:

1 *Thyroid stimulating hormone* (TSH), which increases the activity of the thyroid gland and the production of thyroxine. The production of TSH by the pituitary gland falls off when thyroxine is in excess in the blood and vice versa. TSH thus controls the level of activity of the thyroid gland.

2 *Adrenocorticotropic hormone* (ACTH), which controls the activity of the suprarental cortex (p. 184).
3 *Gonadotropic hormones*, which influence the testis and the ovary (p. 188).

Functions of the posterior lobe

Extracts of the posterior lobe of the pituitary gland contain three separate hormones, *oxytocin*, *vasopressin* and *antidiuretic hormone*. Oxytocin causes contraction of smooth muscle in the gut, the ureter and especially the pregnant uterus. Vasopressin produces constriction of smooth muscle in the walls of arterioles and raises the blood pressure. Antidiuretic hormone (ADH) increases the re-absorption of water from the renal tubules and therefore opposes the diuresis or excretion of fluid which follows a large drink of water. It is part of a sensitive mechanism for the control of the body water and of kidney function (p. 99).

The pineal gland

This tiny gland is about 8 mm long and lies towards the back of the brain in the mid-line. It consists of secretory and neuroglial cells. Its function is obscure but it is possibly concerned in the regulation of reproductive activity.

The thyroid gland

The thyroid gland lies in the lower part of the neck in front of the trachea. It weighs about 25 g and has two oval lobes joined by a narrow band or isthmus. The gland consists of large numbers of follicles, which are spheres of cuboidal epithelial cells surrounding a jelly-like mass called colloid. They secrete *thyroxine*, which is a combination of iodine with amino acids. The normal iodine requirement of the body is 100–200 μg daily, which is absorbed into the blood stream from the intestine. The thyroxine is stored in the colloid of the follicles. Normally about 0.35 mg of thyroxine is released daily, depending on the amount of iodine available and on the activity of the thyroid stimulating hormone (TSH) of the pituitary gland. The production of TSH is directly related to the amount of thyroxine circulating in the blood. Excess of thyroxine in the blood cuts down the output of TSH and deficiency of thyroxine stimulates the pituitary gland to secrete TSH. In this way the requirements of thyroid hormone are accurately met. Thyroid hormone acts as a catalyst for the oxidative reactions of body cells. It regulates the overall metabolism of the body and controls the rate of oxygen consumption and therefore the metabolic rate. It speeds up chemical reactions and releases extra energy and heat. It also causes increased tissue breakdown and thus controls the rate of oxygen consumption. It is important for the regulation of body temperature and is vitally concerned in the proper rate of growth in immature animals. Thyroxine is such a potent agent that excess of it produces a wasting disease – *thyrotoxicosis*. Lack of it causes dwarfism in childhood (*cretinism*) and lethargy in adults (*myxoedema*).

There are also smaller numbers of perifollicular cells, which produce calcitonin, a hormone concerned with calcium metabolism (see below).

Parathyroid glands

There are four parathyroid glands, each weighing about 50 mg, situated in close proximity to the posterior surface of the thyroid gland, but unlike other endocrine glands, they do not seem to be under the control of the pituitary gland. They secrete a hormone, *parathormone*, which is important in the regulation of the metabolism of calcium and phosphorus (Fig. 15.2). The normal person requires 1 g of calcium daily (more during pregnancy) and obtains this from milk, cheese, eggs, beans, nuts and green vegetables. The body contains 1500 g of calcium, 99 per cent of which is in the bones and teeth. The absorption of calcium depends on the diet, the acidity of the gut, the amount of phosphates and fatty acids present and on the presence of vitamin D. The calcium is transported in the serum and the normal serum calcium level is 5 mmol/l (10 mg per cent). Calcium is essential for the function of nerves, muscles and the heart, for the permeability of cell membranes and for the clotting of blood.

The secretion of parathormone by the parathyroid glands and of calcitonin from the thyroid gland is directly related to the amount of calcium ion present in the plasma and tissue fluids. If the level of calcium ion falls too low parathormone is produced but if the level

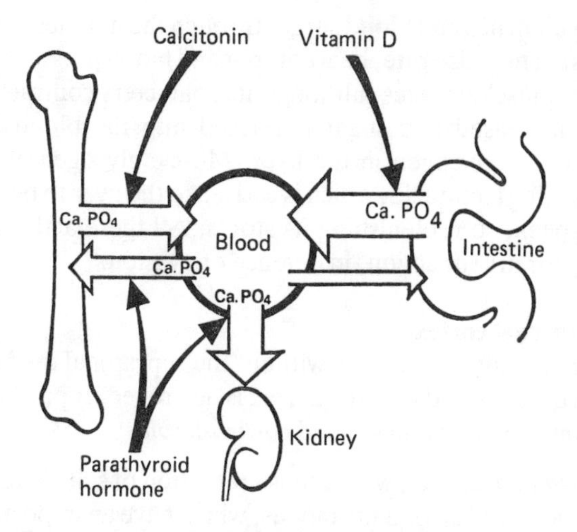

Fig. 15.2 The factors concerned in the regulation of the level of calcium in the blood. Ca: calcium; PO_4: phosphate ions.

rises too high calcitonin is secreted, which lowers the blood calcium level by a direct action on bone, preventing its breakdown. Parathormone increases the number and activity of the osteoclast cells in bone (p. 36), with consequent resorption of bone and the liberation of calcium and phosphorus.

Suprarenal (adrenal) glands

There are two suprarenal glands, perched like cocked hats over the kidneys. They are yellow and weigh 5 g each. Each consists of a thick outer layer (*cortex*) and a small central *medulla*.

The suprarenal medulla

The medulla comprises only 5 per cent of the whole gland and is best regarded as an extension of the sympathetic nervous system. The medulla responds to nerve signals only, by producing a potent hormone, *adrenaline*. This circulates in the blood stream and reproduces or augments all the effects of sympathetic nerve activity (p. 159). The medulla is activated by stress, such as fear, and is responsible for most of the characteristic manifestations of fright.

The skin blanches and blood surges through the muscles of the limbs and heart. The pulse rate, heart output and blood pressure rise. The intestinal muscle relaxes, although its sphincters contract. Metabolism is increased and sugar is poured into the blood from the breakdown of glycogen in the liver. Muscle glycogen also breaks down forming lactic acid, which is carried to the liver to be reformed into glycogen, so replenishing its stores. All these actions prepare the body for urgent action, in defence or evasion.

The suprarenal cortex

The human body can survive without the suprarenal medulla but it fails and eventually dies if the cortex is destroyed. It produces three main groups of hormones called *corticosteroids*:

1 *Mineralocorticoids*, which cause retention of salt (NaCl);
2 *Glucocorticoids*, or S hormones, which have an important effect on sugar metabolism;
3 *Androgens*, or N hormones, which cause retention of nitrogen.

The mineralocorticoid hormones cause retention of sodium and chloride in the body, and excretion of potassium, by their action on the renal tubules. *Aldosterone* is the most important member of this group. The increased amounts of sodium and chloride in the tissue fluid tend to increase its osmotic pressure. This causes, directly in the kidney itself, and indirectly by the antidiuretic hormone mechanism (p. 181), an increased absorption of water from the renal tubules. These compounds decrease muscular fatigue and increase work capacity. They are essential for the maintenance of life.

The glucocorticoid hormones render the cells of the body more resistant to all types of injury. *Cortisone* is a member of this group. They help the body to resist stress, such as cold and injury, and they diminish its reaction to irritants and infections. They also diminish or abolish the hypersensitivity reactions to foreign proteins (p. 91). They interfere with wound healing and antibody formation and patients receiving cortisone may react badly to infections or operations.

The androgen or N hormones lead to the building up of protein and nitrogen retention. They also influence the secondary sexual characteristics in the direction of masculine features and in excess they may produce severe masculinisation in women and children. They are important for the maintenance of the normal body structure, for repair after injury or disease and for growth.

CONTROL OF THE SUPRARENAL CORTEX

The suprarenal cortex is directly controlled by the pituitary gland through the secretion of *adrenocorticotropic hormone* (ACTH). It is believed that the normal reactions to physiological stresses such as cold, injury and infection are mediated through the pituitary gland by the secretion of ACTH, which stimulates the suprarenal cortex to produce the necessary quantities of corticosteroids.

The pancreas

This digestive gland was described on page 79 but it also has an endocrine component formed by two types of cells. There are numerous clumps of specialised cells called the Islets of Langerhans. About 80 per cent of these cells (*B cells*) secrete the hormone *insulin* into the blood stream. The remainder (*A cells*) secrete *glucagon*. The two hormones are vitally concerned in the metabolism of carbohydrates, proteins and fats and in the regulation of the blood sugar level.

Blood sugar regulation

The amount of sugar in the blood stays close to 4.0 mmol/l (80 mg per cent) and even after a meal the level only rises to 8.0 mmol/l (160 mg per cent) for an hour or so in healthy people. Glucose is the only fuel used by the brain cells and accurate control of its level in the blood is therefore essential (Fig. 15.3). It is constantly being absorbed from the intestine or formed from other sources in the body and it is stored in the liver and muscles. Insulin is secreted in response to a rise in the blood glucose level and facilitates the entry of glucose into the cells especially of muscle and liver. If the blood glucose level falls, glucagon is secreted and breaks down glycogen in the liver so that the blood glucose level rises. Insulin increases the uptake of glucose and amino acids by most cells and thus stimulates glycogen, fat and protein synthesis.

The growth hormone of the anterior pituitary gland opposes the action of insulin, reduces the tissue utilisation of glucose, and raises

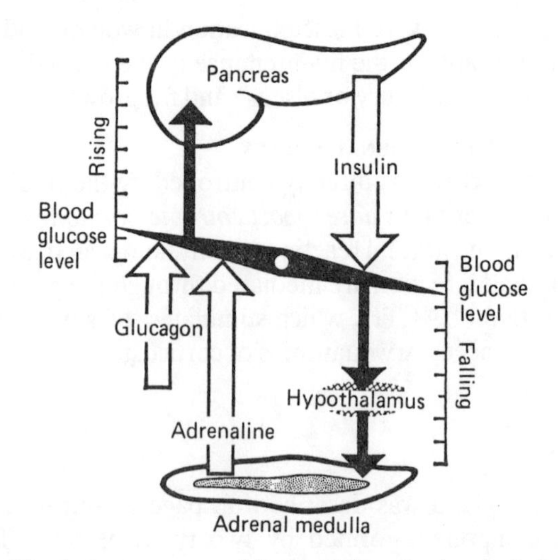

Fig. 15.3 The factors concerned in regulating the level of glucose in the blood

the blood sugar level. It also increases the formation of glycogen in the liver from protein and fat. Adrenaline is secreted in response to pain, fear or anger and increases the breakdown of glycogen to glucose in both liver and muscle, thus preparing the organism for either struggle or escape in an emergency. The secretions of the adrenal cortex, which are steroid compounds like cortisone, have similar actions to the anterior pituitary hormone. Thyroxine also tends to increase the blood sugar level. These mechanisms interact to stabilise the blood glucose level.

16

Reproduction

All living creatures are normally able to reproduce themselves and the mechanisms concerned at the level of the individual cells were described in Chapter 3. In humans two specialised cells, one from each parent, are created and eventually united by the activities of the reproductive organs.

Male reproductive organs (Fig. 16.1)

The male cells, *spermatozoa*, are produced in two glands, the *testes*, which hang in the *scrotum*, below the *penis*. They develop inside the abdomen but descend into the scrotum just before birth. Each testis has a tough fibrous capsule enclosing coils of tubes, which produce the spermatozoa, and interstitial cells, which secrete a hormone called *testosterone* into the blood. The many tubules of each testis join to form the *epididymis*, which passes into a single duct on each side. These *spermatic ducts* run up over the brim of the pelvis and pass down to the base of the bladder, where they penetrate the *prostate gland* and finally open into the *urethra* (p. 188). There is a small reservoir, or *seminal vesicle*, for storing spermatozoa attached to each duct just behind the prostate.

Spermatozoa are minute, highly motile cells with a small head, containing nuclear material (DNA), and a long, active, sinuous tail. They are formed in enormous numbers from puberty onwards and several hundred million of them are expelled on every sexual occasion. Normally, *semen*, which is prostatic secretion swarming with spermatozoa, passes along the urethra of the enlarged, erect

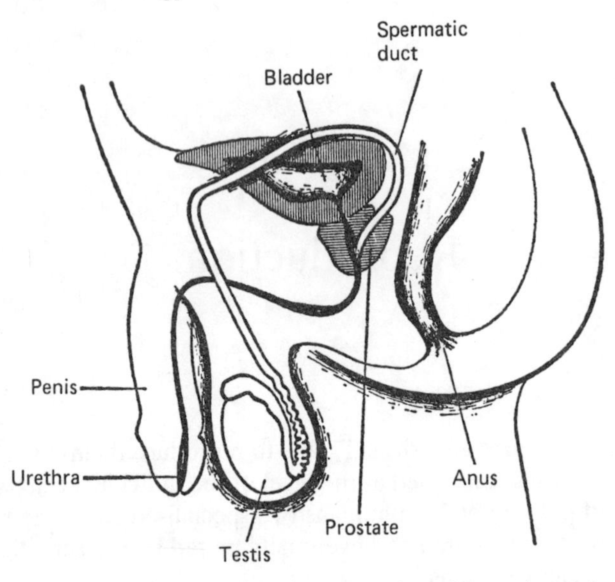

Fig. 16.1 Diagrammatic cross-section of the male pelvis

penis and is deposited in the vagina. The spermatozoa travel actively until one enters a female ovum and conception occurs.

The penis consists of three elongated masses of vascular, spongy tissue. One of them contains the urethra, which opens at the external meatus. During erection the spongy tissue becomes enlarged and stiffened by the entry of blood.

The hormone, testosterone, produces the characteristic male sexual features. These are the masculine shape, bone structure and muscular development, hair distribution and genital size. The secretion of testosterone is controlled by a hormone released from the pituitary gland. It follows that inadequacy of either the pituitary gland or the testes may produce sexual immaturity.

Female reproductive organs

There are two pinkish-grey, almond shaped *ovaries* 3 cm × 1.5 cm × 1 cm. They produce germ cells – *ova* – and female sex hormones. An ovum is released every month throughout the period of sexual maturity from puberty to the menopause. The two ovaries are situated one at each side of the lower part of the abdominal cavity, at the brim of the pelvis (Fig. 16.2). Each one is close to the opening of one of the two *Fallopian tubes*, which lead into the *uterus*. The uterus is a hollow organ, 7.5 cm long, with a thick muscular wall and a lining of mucous membrane with a rich blood supply – the *endometrium*. The cavity of the uterus communicates with the thin-walled *vagina* by a narrow canal running through the *cervix*. The vagina terminates in the *vulva*. This consists of fleshy lips – the *labia* – and also contains the exit of the urethra and the sensitive *clitoris*.

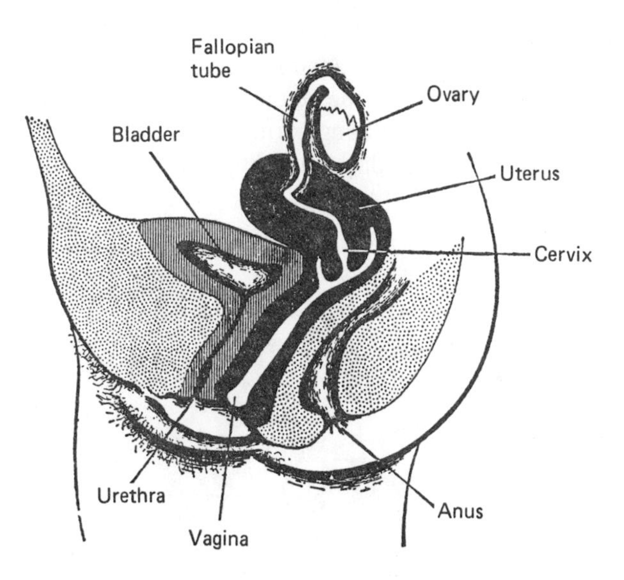

Fig. 16.2 Diagrammatic cross-section of the female pelvis

At ovulation each month one or other ovary releases an ovum, which passes along the Fallopian tube to the uterus. On a propitious occasion one spermatozoon succeeds in penetrating far enough to fertilise it, usually in the Fallopian tube and the fertilised ovum attaches itself to the inner wall of the uterus. It develops rapidly into an embryo and at three months it is a foetus with some human features. It is finally expelled through the vagina as a newborn infant, ten lunar months (280 days) after conception.

The female reproductive cycle

Extremely elaborate structural and hormonal changes occur regularly throughout the period of sexual maturity in the female. The primitive cells which later develop into ova are present in the ovary from birth. At the age of 11–14 years the anterior pituitary gland begins to secrete *follicle stimulating hormone*, which starts off characteristic changes in the spherical masses of cells called the ovarian *follicles*. Each follicle begins to enlarge and becomes fluid-filled with a centrally placed viable ovum (Fig. 16.3). The follicle secretes another hormone, called *oestrogen*, which has two actions. It stimulates the lining of the uterus to become thicker, with more glands and blood vessels, ready to receive the ovum if fertilisation occurs. It also acts upon the pituitary gland itself and stops the secretion of follicle stimulating hormone.

On average, fourteen days after the start of the cycle the follicle

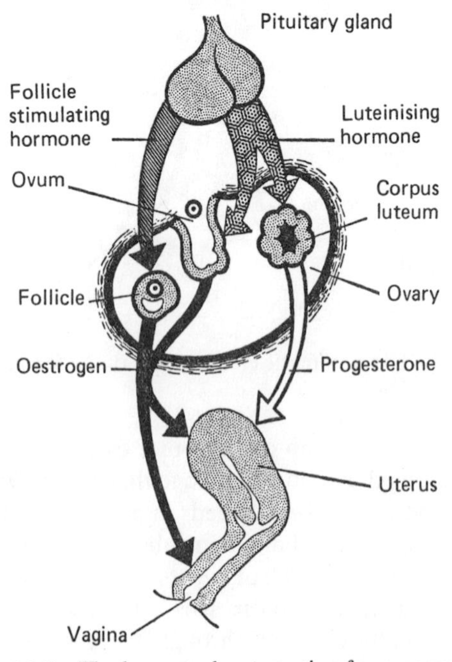

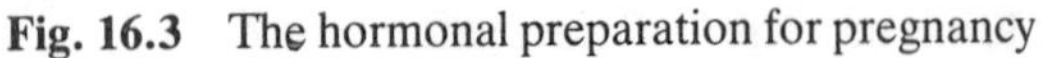

Fig. 16.3 The hormonal preparation for pregnancy

bursts and the ovum starts its journey to the uterus. The ruptured follicle now develops, under the influence of a further pituitary hormone, *luteinising hormone*, into a yellow mass of cells in the ovary called a *corpus luteum*. This secretes *progesterone*, which greatly increases the preparative changes in the uterine lining and causes supplies of nutrient material to collect there (Fig. 16.4). If pregnancy occurs, the corpus luteum continues to produce progesterone during the first three months. If the ovum is not fertilised, the corpus luteum degenerates on about the twenty-eighth day of the menstrual cycle and the supply of oestrogen and progesterone is cut off. The blood vessels of the lining of the uterus constrict causing local ischaemia and eventual death of the inner layers. They are discharged through the cervix with loss of blood, which continues for several days and forms the menstrual period. The pituitary gland at once starts to produce follicle stimulating hormone again and the cycle of ovulation (preparation for conception – menstruation)

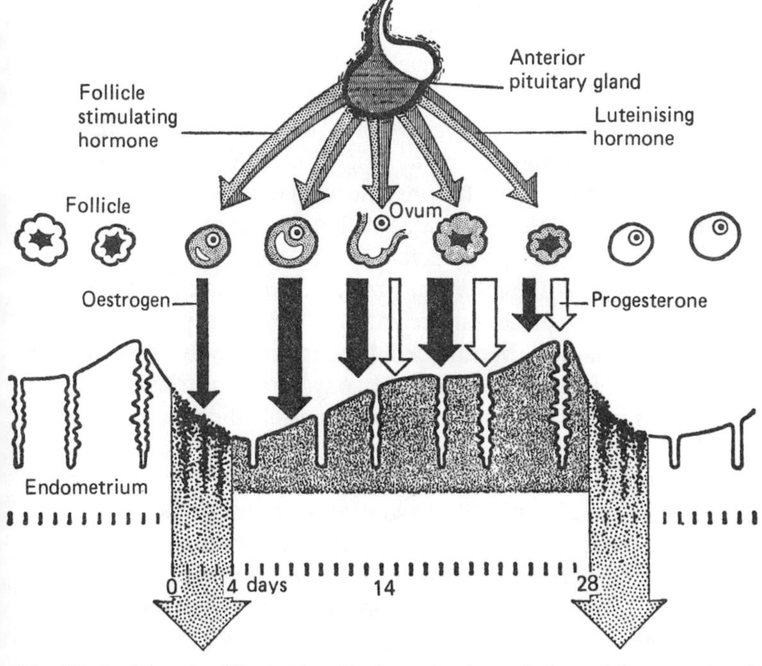

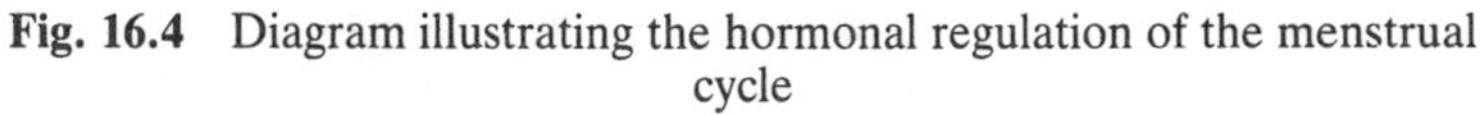

Fig. 16.4 Diagram illustrating the hormonal regulation of the menstrual cycle

repeats itself. In normal women this cycle recurs regularly every twenty-eight days until all the available ova have been shed. This usually happens at the age of 45–50 years, when menstruation stops, the woman is no longer fertile and the *menopause* has occurred.

In the event of pregnancy the corpus luteum persists and grows larger so that the production of progesterone is much increased (Fig. 16.4). This suppresses further ovulation, continues the development of the lining of the uterus, prevents its muscular wall from contracting as its contents increase in size and prepares the breasts for the eventual production of milk – *lactation*. If fertilisation does not occur the whole cycle repeats from the first day of the next menstrual cycle.

Pregnancy and childbirth

Fertilisation usually occurs while the ovum is passing along the Fallopian tube and the fertilised ovum becomes implanted in the uterine wall two or three days later. It then begins to divide repeatedly until the fully developed infant is finally ready for delivery nearly 300 days later. The outer layers of the dividing ovum develop into the *amniotic membrane*, which forms a bag filling the uterus and enclosing the foetus (Fig. 16.5). At first the ovum is nourished directly from the uterine wall but soon an elaborate vascular structure, the *placenta*, develops at one part of the uterine wall. The placenta is connected to the vascular system of the growing foetus by the *umbilical cord*, which contains two arteries and one vein. The placental circulation is entirely separate from the mother's circulation, but the cellular partitions between them are exceedingly thin and free diffusion of oxygen and nutrients into, and of carbon dioxide and waste products out of, the placenta readily occurs.

The placenta also produces large amounts of three hormones, *chorionic gonadotropin*, oestrogen and progesterone. Chorionic gonadotropin is responsible for the persistence of the corpus luteum during the early part of pregnancy and also forms the basis of the pregnancy tests. Oestrogens and progesterone act together to prevent further ovulation and menstruation during pregnancy. They are responsible for profound modifications in the whole organism which accompany pregnancy. The endometrium of the uterus de-

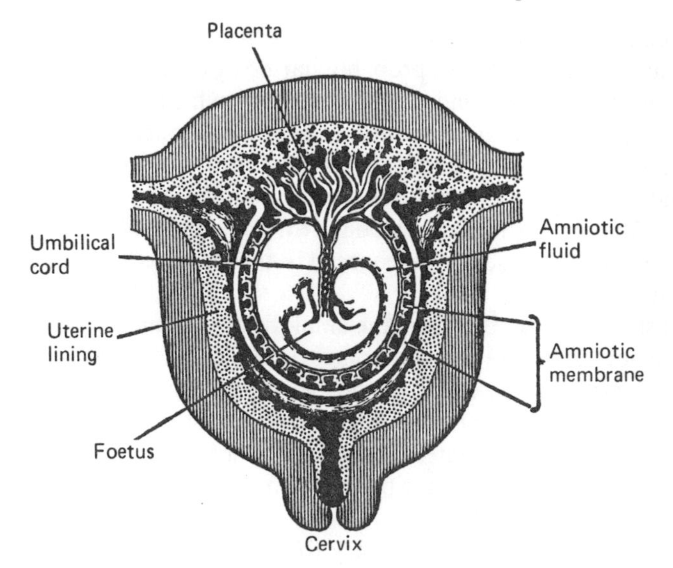

Fig. 16.5 Diagrammatic cross-section of an embryo growing in the uterus

velops further vascular and glandular changes, the uterine muscle relaxes and its cells enlarge 20–100 times, the breasts enlarge and prepare for lactation.

The developing foetus floats in 0.5–2 litres of clear or milky amniotic fluid, which is essential to equalise pressure and stabilise the temperature. The foetal heart can be heard beating at a rate of 120–130 per minute from about the fifteenth week of pregnancy. Movements of the limbs can be felt now and the foetus has periods of sleep, rest and movement. The normal infant at term is 50 cm long and weighs about 3.5 kg. The foetal lungs are inactive and are excluded largely from the circulation until the last month of pregnancy. After delivery important rearrangements of the circulation occur (p. 120).

After 280–300 days the amniotic sac bursts, the uterus begins to contract rhythmically and eventually the baby is born, followed by the placenta. During the next 6–8 weeks the uterus contracts to about normal size. Ovulation and menstruation may reappear two to six months after delivery even if lactation continues. It is quite possible for a woman to conceive again while lactating.

Childbirth (parturition)

The onset of parturition is probably started by the foetus producing hormones (ACTH and corticosteroids) which influence the placenta to reduce the secretion of progesterone and produce prostaglandins, which are fatty acids that constrict blood vessels and stimulate muscle. This induces uterine contraction and begins to stretch the cervix (Stage 1 of labour). Stretch receptors in the cervix induce reflex autonomic responses via the spinal cord giving enhanced contractions.

At some point in this stage the amniotic membranes rupture and fluid is released. The cervix becomes fully dilated (Stage 1) and the foetal head progresses down the pelvic canal and is eventually delivered (Stage 2). During Stage 1 the bag of fluid protects the foetal head and eliminates any danger of infection from the maternal passages. Stage 3 is the period from the delivery of the baby to the discharge of the placenta.

There is an increasing tendency to use sophisticated electronic techniques during labour to monitor the condition of mother and baby. This enables appropriate measures to be applied if distress should appear.

Lactation

The breasts are rudimentary in both sexes until puberty, when the female breast begins to enlarge during each menstrual cycle under the influence of oestrogen and progesterone. The active tissue consists of alveoli, lined with epithelium, which are connected by ducts to the nipple. The enlargement after puberty partly involves the alveoli, but there is also a laying down of fat.

During the first half of pregnancy the oestrogen and progesterone secreted by the placenta, plus hormones from the anterior pituitary and thyroid glands, cause a marked overgrowth of the alveoli and ducts. Later in pregnancy the blood flow through the breast increases, the epithelial cells swell up, milk begins to form and the breasts become tense. The nipple enlarges, becoming tense and erect, and the areola around it deepens in colour.

Immediately after delivery a thick yellow fluid called *colostrum*, which is rich in protein and salts but poor in carbohydrates and fat, is produced in small quantities. After a few days a free flow of milk can be obtained from the breasts and further secretion of milk depends on the production by the anterior pituitary gland of a hormone called *prolactin*. Prolactin secretion is a reflex response to the stimulation of suckling and the more vigorous the suckling the better the supply of milk. The discharge of milk from the breast depends partly on the baby's suction and partly on the stimulation of the duct walls by *oxytocin*, which is produced reflexly from the posterior pituitary gland by the stimulation of suckling.

Infant feeding

Human milk is exactly suited to the needs of the baby and increases in nutritional value up to the end of the first month, when the infant's digestion is better developed. It differs in composition from cow's milk in having less protein and more sugar (Table 16.1). In addition to immunoglobulins essential for early defence against infection, milk also contains small quantities of salts such as the phosphates and chlorides of sodium, potassium and calcium, plus vitamins A, D and C in amounts depending on the mother's diet.

	Protein	*Lactose*	*Fat*
Human milk	1.5	7.0	3.0
Cow's milk	3.5	4.5	3.5

Table 16.1 The relative compositions of human and cow's milk in grams per 100 ml

At birth the digestion and absorption of nutrients is less efficient than in the adult and up to 30 per cent of the energy value in bottle-fed infants may be lost in the stools. This is largely due to defective fat absorption resulting from inadequate production of bile salts and pancreatic lipase. During the first few days after delivery the breasts secrete colostrum, a thin yellow fluid with a high protein, low fat content. This slowly changes over to normal breast milk and the digestive efficiency improves. Weaning to solid food should occur gradually between the ages of six to ten months. The normal infant regains its birth weight at ten to fourteen days. It then gains about 30 grams a day. A baby weighing 15 kilograms at birth should weight about 30 kg at six months and 45 kg at one year.

Breast feeding is ideal because the milk from a healthy mother is unique in quality and quantity. It is maintained even when the mother is suffering from starvation or disease. The act of breast feeding is of great emotional significance and ensures proper 'bonding' between mother and child. Human milk bestows passive immunity from infection through immunoglobulins which act directly and by changing the pH of the stools. It also favours the growth of harmless lactobacilli which reduce the growth of colonic bacteria. There is no danger of excess amino acids damaging the infantile nervous system and there is less obesity in breast-fed infants. Disease and 'cot deaths' are less common in the breast fed.

If breast feeding is impossible or inadequate, bottle feeding with modified cows's milk or powdered milk is substituted. The protein of cow's milk is largely caseinogen, which is scanty in human milk. It tends to coagulate in the human stomach forming large insoluble masses and this is one of the reasons why cow's milk has to be diluted and modified before being fed to infants. Careful attention to sterilisation and cleanliness is essential to avoid infection. If powdered milk is not reconstituted very accurately as directed there can be serious problems from excess salt intake causing fits, coma or death.

Boy or girl?

The intricate mechanisms controlling heredity were described in Chapter 3. The controlling influence is the deoxyribonucleic acid (DNA) in the chromosomes in the cell nucleus. The chromosomes responsible for sex have been labelled XY for male and XX for female (Fig. 16.6). All ova contain X chromosomes only, but spermatozoa can contain either X or Y chromosomes. If an X-carrying spermatozoon meets the X ovum, the result is a girl, but a Y-carrying spermatozoon always produces a boy. Most men produce approximately equal numbers of X and Y spermatozoa and there is equal probability that the offspring may be of either sex.

Twins

There are two kinds of twins. Occasionally two ova are produced at a single ovulation and both are fertilised. The genetic pattern of the two is not the same and *non-identical twins* result. Sometimes a defect occurs in the development of the ovum after fertilisation. It splits in half and two individuals develop with exactly the same genetic constitution. These are *identical twins*.

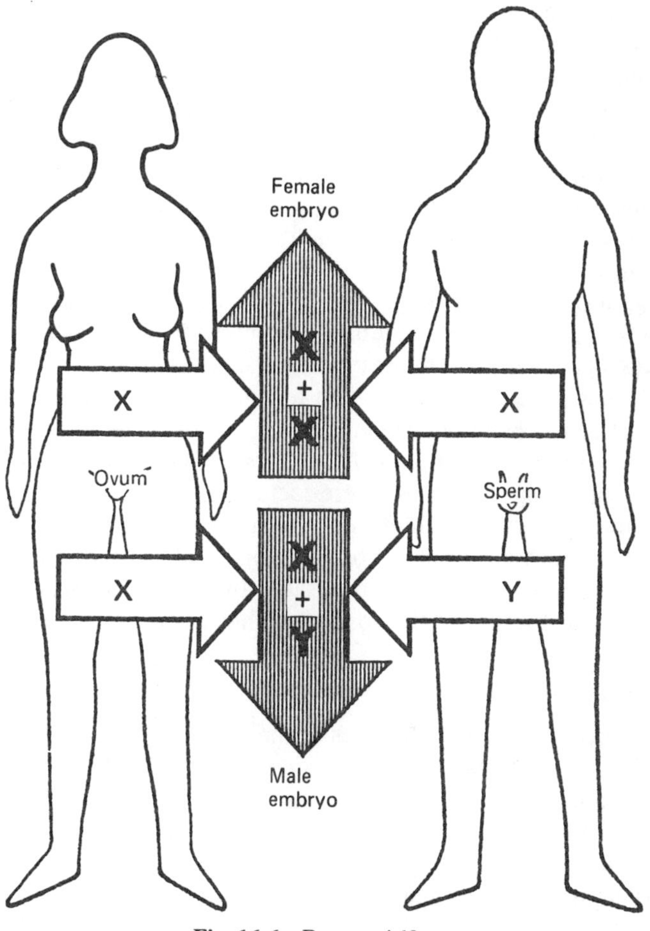

Fig. 16.6 Boy or girl?

Pre-natal testing

Regular antenatal supervision has greatly reduced the risks of pregnancy to mother and child. Any obvious physical malformation, such as a narrow pelvis, can be noted and complications avoided by early induction of labour or even surgical delivery by Caesarian section.

The mother is weighed regularly to monitor foetal development, maternal nutrition and possible water retention. The size of the uterus is checked regularly in relation to the stage of pregnancy and to detect abnormalities of the foetus or excess of amniotic fluid. The blood pressure is recorded because certain complications are heralded by a rise in blood pressure. The combination of high blood pressure, oedema and albumin in the urine is called *pre-eclampsia*. It may lead on to premature separation of the placenta with damage or even death of the child or its mother. Rarely pre-eclampsia is followed by eclampsia with coma, fits and, if treatment fails, death.

The urine is tested for protein to warn of developing eclampsia or renal impairment. If sugar is present it may be lactose from the breasts but if diabetes mellitus is present, with glucose in the urine, special care is needed to preserve mother and child.

A specimen of blood is taken to establish the presence of any anaemia, to check the blood group (p. 85) in case transfusion is necessary and to identify the Rhesus state of the mother (p. 86).

In certain special cases where foetal abnormality is suspected a specimen of amniotic fluid can be aspirated through a needle and the cells checked for chromosomal abnormalities such as Down's syndrome (p. 28). This is facilitated by ultrasound scanning to identify the foetus and the position of the placenta.

Population control

The world population increases by many millions every year and is beginning to make intolerable demands on food supplies and resources of fuel and raw materials. The population of individual countries may rise and fall independently of the overall changes and the reasons are not fully understood. In developed countries artificial limitation of pregnancy is widely practised.

There are three main methods of contraception available:

1 *Mechanical and/or chemical* Various mechanical devices can be used to prevent the union of spermatozoa and ovum. They are usually made of rubber or plastic and are used by one or other partner. Chemicals that destroy or immobilise the spermatozoa are also available. None are completely reliable. Devices of plastic or metal can be inserted into the uterus (intra-uterine devices, IUD) for long periods. They interfere with the implantation of the embryo in the wall of the uterus, preventing pregnancy.

2 *Oral* Various combinations of the hormones oestrogen and progesterone can be taken by mouth in such a way that the process of ovulation is inhibited, although menstruation occurs as usual. These compounds are very reliable contraceptives, but they have occasional side effects, some of which are dangerous.
3 *Surgical* In men the spermatic duct can be severed surgically and the spermatozoa can no longer be ejaculated. The operation is safe and minor: sexual capacity is unaffected. In women the Fallopian tubes can be tied permanently, preventing access of spermatozoa to the ovum.

17

The Environment

Human beings have evolved a bewildering array of defences and adjustments which enable them to survive in spite of a multitude of adverse environmental factors. These are either physical or biological. Adequate supplies of air, food, water and sunlight are essential as is protection from climatic extremes. Small groups living in primitive conditions can survive and even flourish by their own efforts and the observation of simple rules and rituals. Larger populations have to develop elaborate controls of the physical and biological environment in order to survive. The larger the population group the greater the pressure it places on the environment and the greater the care and regulation required.

Climatic factors

Temperature

Adequately fed, sheltered and clad we can tolerate a wide range of temperatures from well below 0 °C to nearly 100 °C. The mechanisms are fully described on pages 64–6. Acclimatisation to very high or low temperatures does occur and Australian aborigines were found who could tolerate subzero temperatures while lying naked on the ground all night. Prolonged exposure to excessive heat, especially if humidity is high, can lead to heat exhaustion with prostration and a fall in blood pressure which is usually recoverable. If the corrective mechanisms fail, the body temperature rises to the point of heat stroke, which may lead to coma, irreversible damage to the brain, heart, kidneys and liver and even to death.

The converse condition of excessive cooling or hypothermia is a special problem in infants and the aged. The newborn, especially if premature, has little capacity for heat production when cooled. It also has a large body surface area in relation to its mass and loses heat rapidly if not carefully protected. Over the age of 65 body temperature can easily fall below 35 °C from causes such as exposure, lack of fuel, poor nutrition and various diseases and endocrine deficiencies. The poor and elderly living alone need careful supervision and assistance.

There are two forms of local cold injury. Prolonged exposure in cold water causes *immersion foot* with damage to nerves, blood vessels and muscle due largely to lack of oxygen. This is potentially reversible if the foot is carefully rewarmed. Frostbite causes permanent damage to the skin and deeper tissues with gangrene because of disruption by ice-crystal formation as well as loss of blood and oxygen supply.

Air

Man is very sensitive to the oxygen content of the air. Up to 20 per cent of the world population live at altitudes of 2000 metres (6500 feet) and are satisfactorily acclimatised, largely by an increase in the haemoglobin content of the blood. Permanent residence is possible up to about 5000 metres. Special training has made it possible to climb Mount Everest at over 8500 metres, but the risk of permanent brain damage is high. Rapid ascent to height often causes mountain sickness, with headache, breathlessness, fatigue and mental confusion. In more severe cases the lungs become water-logged (pulmonary oedema) and death may follow.

Exposure to high air pressure as in diving also has many risks. It is possible to withstand a pressure equivalent to a depth of 600 metres (2000 feet) using special techniques. At high pressures gases are forced into the blood and if subsequent decompression is too fast severe damage to the nervous system, lungs, ears and bones may occur. At high atmospheric pressure nitrogen becomes poisonous and may cause irrational behaviour or even coma; an effect similar to that of ethanol. Oxygen under pressure may also damage the nervous system.

Air pollution

Atmospheric pollution is the indirect result of the concentration of people into urban communities which have grown steadily larger. Its dangers are well recognised and most developed countries have introduced extensive legislation and inspection to control it. The use of smokeless fuel and the improved design of grates and engines has produced a marked improvement in the atmosphere in the last two decades. Smoke Control Areas have been established and are monitored by the Local Authorities in Great Britain. Industrial processes are closely controlled by the Health and Safety Executive (1974) and by the Local Authorities. The effect of these measures is shown by a reduction of the average amount of smoke in London by two-thirds in the last twenty-five years and an increase in recorded winter sunshine from 1.2 hours/day to 1.8 hours/day.

Eighty per cent of the pollutants consists of smoke from domestic sources and 20 per cent are gas and vapour from industrial activity. Ionising radiation from nuclear power stations and weapon testing is a separate and controversial problem. Air pollution reduces the amount of ultraviolet radiation from the sun reaching ground level. It also leads to an increase in fog formation, retards the growth of vegetation and damages buildings and fabrics. The irritant gases cause increased morbidity and mortality from chronic bronchitis and possibly, in the long term, lung cancer.

Road transport exhaust fumes produce carbon monoxide and carbon dioxide, oxides of sulphur and nitrogen, unburnt hydrocarbons and lead. Children are specially at risk because lead is a poison which affects the nervous system. There is no clear evidence of a safe level for lead and the effect on the intelligence quotient (IQ) of children has been shown to be dose dependent. No urban children are entirely free from lead but there are many sources apart from transport. Active steps are being taken to reduce the lead content of petrol.

Radiation

There has always been natural background radiation from the earth's crust and cosmic rays amounting to about 100 mrem/year per person. In recent years we have added about 20 mrem/year from medical sources such as X-rays and isotopes. Industry, weapon testing and nuclear power add a further 10 mrem/year. The total exposure is normally less than 150 mrem/year per person. Rem means roentgen equivalent man and establishes a biological equivalent with the effects of X-rays.

The biological effects of radiation depend upon the dose and the duration of exposure. There is probably no 'safety level' even at very low doses. Exposure to high doses causes illness or death within days. Survivors have an increased risk of developing leukaemia or cancer of various organs. In pregnancy there is a danger of death or malformation of the foetus and leukaemia may develop later. In addition there is an increase in foetal abnormalities, abortion and sterility in later years.

There are comprehensive regulations for the control of radiation under the general supervision of the National Radiation Protection Board. At all levels there is careful supervision of environmental and personal exposure, the transport of radio-active materials, the management of radio-active waste and the control of any accident hazards.

Water

There is an abundant supply of water in Britain resulting from an average rainfall of 50–100 cm (20–40 in) per year. Its collection, storage and delivery to domestic and industrial consumers is the responsibility of nine Regional Water Authorities. The purity of water is also controlled by the environmental health officers of the Local Authorities.

Surface water in lakes, reservoirs and rivers is an important source but is liable to contamination. Rainwater penetrates the surface and percolates through to form huge reserves of underground water resting on rock or impermeable clay. This water is available from springs and shallow wells. Deep wells produce a supply of pure water very suitable for large populations as in London.

The water supply is checked regularly for chemical and bacteriological content. Chemical analysis shows the presence of excess of toxic substances such as lead or arsenic. High nitrate and ammonia levels suggest animal contamination. The routine bacteriological test is for the presence of coliform organisms which indicate recent faecal contamination either human or animal. Many diseases are associated with impure water. Bacterial infection induces typhoid, cholera and dysentery while viruses cause infective hepatitis and poliomyelitis. Tropical parasitic diseases such as amoebiasis, filariasis, bilharzia and hookworm are also waterborne.

Water can be purified in various ways. Storage in reservoirs for 1–2 weeks allows sediment to settle with a reduction in the amounts of bacteria and solid matter. Filtration through sand beds or artificial filters leads to further improvement. In many cases chlorination by liquefied chlorine gas is needed to produce safe water, even at the expense of palatability. On a small scale the safest method is boiling for an adequate period.

There has been considerable controversy over the addition of fluorine to the public water supply. It is known that areas in which the water supply contains more than 1 part per million of fluorine have a very low rate of dental caries amongst the children who live there. A number of Area Health Authorities have requested their Regional Water Boards to add fluorine to the water supply to bring its level up to 1 part per million. This is claimed to have significantly reduced the amount of dental caries with resulting personal and economic benefit. This action has, however, been widely criticised on the grounds of intrusion into public rights and of a possible risk of poisoning. The practice of adding fluorine to the water supply is slowly spreading in Great Britain.

Food

The basic food requirements in terms of energy and essential constituents are described in Chapter 6. An adequate supply of wholesome food must be available for consumers throughout the country. This involves the prevention of deterioration or damage by bacterial invasion or enzyme activity.

Food can be preserved in various ways. Fish, meat, fruit and vegetables can be precooked to ensure sterility and then preserved by cold which inhibits bacterial growth. Rapid freezing to −25 °C retains both quality and flavour. The frozen food can then be stored at −10 °C safely for long periods. Frozen food must be given adequate time to thaw before conventional cooking. This ensures complete penetration by the heat with the elimination of any residual infection. Food can also be preserved for shorter periods by chilling at 0 °C. Other methods are high-temperature-heating in tins, dehydration, smoking, salting, pickling and irradiation, but they all change the palatability, appearance and utility of the material. Chemicals such as nitrites in meat products, or sulphur dioxide and benzoic acid in preserved fruits, can also be used but there are potential dangers such as the induction of cancer.

Certain foods present special problems. Milk is easily adulterated, deteriorates rapidly and can transmit infection. Adulteration results from the addition of water, skimmed milk, thickeners and chemical preservatives. This can be controlled by laboratory testing. Bacteria flourish in milk and tuberculosis, diphtheria, brucellosis and coccal infections can all spread easily from infected cows. Milk can also be contaminated by typhoid and dysentery-causing organisms from careless milk handlers or contaminated water. Careful inspection and regulation of the cattle, milking premises and personnel is fundamentally important. Most milk is transported in bulk and treated before bottling in special registered premises. It can be rendered safe by pasteurisation or sterilisation. Pasteurisation is used for 95 per cent of the milk consumed in Britain. It consists of heating fresh, clean milk for varying periods at 60–72 °C before bottling. It minimises the danger of infection but it does not affect the nutrient status of the milk. Sterilisation involves heating at 100 °C in the bottle. It is safe but may affect the nutrient status. Other forms of heat treatment produce safe milk which can be stored for long periods in bottles or special containers.

Housing

Housing is an important factor in the health and social stability of a community. Bad housing spreads infection and impairs social standards. Probably 10–15 per cent of the housing stock in Great Britain is unfit or substandard. The local authorities are responsible for the supervision of housing and for its improvement. In spite of great efforts the situation is still serious for financial and social reasons.

In some areas virtually the whole of the housing stock may be defective, constituting a slum. In other areas, although the bulk of the housing is satisfactory, individual houses may have been allowed to deteriorate. The passage of time and changing social patterns have accentuated the long-standing housing shortage.

Adequate housing must be well maintained, dry, well lit and properly ventilated. Hot and cold water, a sink, internal toilet, a bath, proper drainage, gas or electric supply, adequate heating and facilities for the storage, preparation and cooking of food must be available.

Noise

The harmful effects of noise are now well recognised. In addition to emotional distress and disturbed sleep, permanent deafness with failure of response to higher frequencies is common. Such hearing loss may be a severe disability, especially in noisy surroundings and cannot usually be corrected by the use of hearing aids.

Noise arises from transport and recreation on land, sea and in the air. Industrial noise can be a serious problem to workers in factories and foundries. Recreational noise has been greatly increased by the use of electronic amplifiers. Gunfire is a significant source of noise both from military activities, even in peace time, and from sporting use.

Noise can be reduced at source if cooperation can be obtained. Ear protectors should be worn wherever possible. Sound insulation of houses and flats can be helpful in many circumstances.

Waste disposal

Waste may be wet or dry. Dry waste is ordinary household waste from dustbins and the equivalent from industrial sources. Domestic waste is removed regularly by the local authority. If neglected it causes a nuisance from smell, breeds flies and other insects and encourages rats. Industrial waste presents special problems.

Dry waste is usually disposed of by controlled tipping on to sites chosen, often for reclamation. Layers of waste up to two metres thick are covered by earth. The temperature rises to 70 °C and the waste breaks down and is eventually assimilated into the soil.

Some waste is pulverised and used as fertiliser. About 10 per cent of waste is incinerated after separation of salvage, producing useful heat and clinker for road foundations. Dumping at sea is now seldom used.

Industry produces large amounts of solid waste from mines and quarries, forming huge, ugly, spoil heaps. There is also some highly toxic waste which has to be treated chemically or dumped on specially chosen safe tips. Dumping into rivers or the sea may be dangerous.

Sewage is water with organic and non-organic solids. It comes from surface drains and from household and commercial sources. It takes up large amounts of oxygen and is lethal to plant and animal life if discharged untreated into rivers. Initial treatment consists of settlement of solids in sedimentation tanks. The effluent is allowed to percolate through clinker beds where it oxidises. Larger quantities are mechanically aerated over biologically active sludge in tanks and then stored in settlement tanks. The effluent can safely pass into rivers or the sea and the sludge is a valuable fertiliser.

Micro-organisms and disease

Biological dangers in the environment are a wide range of parasites, fungi, bacteria, viruses and other micro-organisms, plus the pests which often transmit them. The great majority of these are essential members of the ecological system and without them life would be impossible. Only a relatively small number are able to affect man. Parasites have been mentioned on page 6. Fungi, moulds and yeasts are widespread and may affect the skin or mucous membranes, often causing diseases such as tinea pedia (athlete's foot), ring worm or candidiasis (thrush). Occasionally they invade the body with fatal results.

Bacteria are single-celled plant-like organisms found almost universally and essential to life. Only a few species are dangerous to man. Every surface of the body, inside and out, is swarming with bacteria and most of them are harmless or normally not invasive. Some of them may suddenly invade the body and cause disease for no obvious reason or because of some local or general change of conditions. There is a natural immunity of some degree to most infections, but this may break down or prove ineffective against a heavy infection.

Viruses and rickettsia are micro-organisms smaller than bacteria which can reproduce only in living cells but can survive for long periods outside them. They all contain the DNA or RNA needed to enable them to multiply utilising the host cell metabolic processes. They produce many different diseases but can also lie dormant in the body for many years. They usually invoke vigorous immunological defense mechanisms, which enables effective vaccination to be undertaken against smallpox, poliomyelitis, mumps, measles, influenza and yellow fever. There is also limited chemotherapy available against viruses in contrast to the abundance of antibacterial agents. Amantadine can ward off influenza and idoxuridine is useful in the treatment of herpes zoster (shingles).

When a virus enters a cell it induces the production of a special protein – *interferon* – which can enter a nearby cell and protect it from virus attack by interacting with cell surface receptors and blocking intracellular virus reproduction. Interferons are effective against some viruses and are also being tried in the treatment of certain forms of cancer. They can now be produced in adequate amounts by special bacterial culture methods.

The pattern of communicable disease is continually varying and many of them are now better controlled than ever before. Smallpox has been eliminated world wide, poliomyelitis has been virtually abolished in Western countries and malaria is partly controlled. Many factors have contributed to this progress. Improved personal hygiene with regular washing and cleansing of hands, body, hair and clothes has become normal because of the greatly improved standards of clothing, housing, water supply, drainage and education. Equally vital have been the advances in engineering, control of insect vectors, immunisation and chemotherapy. In spite of these measures, mosquito-borne diseases are still prevalent. Much has been achieved by the eradication of their breeding places in stagnant water and the use of chemicals such as DDT. Even so malaria is still widespread and debilitating, while dengue, filariasis, yellow fever and a form of encephalitis are also transmitted by these insects.

Insects such as flies carry infection on their bodies and appendages. Food poisoning may be conveyed by flies and their eggs may be deposited rendering food unfit for consumption. Bubonic plague is transmitted from the host rodent to man by the bite of a flea.

Typhus is also carried by fleas or lice depending on the type of infection.

The chief animal pests are brown and black rats, which in addition to their fleas carry salmonella infections (enteritis) and haemorrhagic jaundice. They can be controlled by poisoning, fumigation and proper storage of food with clearance of rubbish. Mice are also prevalent and can carry a form of meningitis. They are controlled in the same way as rats. Pigeons and some cage birds may transmit psittacosis or dysentery-causing organisms.

Cockroaches are very common in warm buildings where food is available. They are hard to control and scrupulous cleanliness and food protection are essential. Bed-bugs have become less of a problem as living standards have improved.

Meat

In Great Britain and most European and American countries meat is carefully inspected after slaughter, which means that direct infection is uncommon. There is a possibility of contamination by tapeworms, tuberculosis, anthrax and actinomycosis. Adequate cooking with prompt consumption avoids most dangers. Meat and meat products may be a source of food poisoning, or even typhoid fever, if they are carelessly handled or stored.

Poultry

There has been a huge increase in the supply of poultry, usually frozen, for human consumption. There is a high risk of food poisoning, which can be avoided by careful thawing, thorough cooking and fast cooling. Poultry must be properly reheated before eating.

T.Y.H.B.—12

18

From Cradle to Grave

As soon as the baby is born, profound structural and functional changes occur, greater than at any other time from conception to senility. It has to adapt to the needs of its separate, although utterly dependent and sheltered existence. These changes include a radical reorganisation of cardio-respiratory function, modification of gastro-intestinal and renal function, rapid colonisation by a variety of bacteria and the acceptance of a sudden flood of sensory stimuli. Rapid physiological change continues throughout the first few months of life and at the end of the first year the infant is in most respects functionally equivalent to the adult.

The newborn infant

The two most important changes are in breathing and the circulation. Previously all exchange of food, gases and waste products has taken place through the placenta, linked to the foetus by the umbilical arteries and vein. At birth the umbilical cord and vessels are cut and within a minute breathing begins. The first breaths are weak and irregular, but soon regular breathing is established. Until now the resistance of the pulmonary arteries is high and their blood pressure is greater than in the aorta. Blood therefore flows from the pulmonary artery to the aorta through the *ductus arteriosus*. There is little flow through the lungs and most of the blood returning from the body passes from the right atrium directly through the *foramen ovale* into the left atrium and on into the aorta.

As the lungs expand with the first breaths there is a sudden fall in pulmonary vascular resistance and a big increase in pulmonary blood flow. Pressure rises in the left atrium and the foramen ovale closes. At the same time pulmonary artery pressure falls below aortic pressure and blood flow in the ductus arteriosus reverses its direction. Aortic blood with a high oxygen content causes constriction of the ductus arteriosus, which closes off completely in the first two or three days after birth. Pulmonary artery pressure and blood flow through the lungs reach adult levels after three or four weeks and the normal pattern of circulation is then established. The systemic blood pressure rises from 80/45 mmHg at birth to 95/65 mmHg at one year and then gradually increases to normal adult values of about 120/70 mmHg at puberty.

In the uterus the infant benefits from a controlled, stable environment and at birth suffers a profound change. Wet and naked, with a large surface area, there is a sudden loss of heat in cool surroundings. There is a rapid increase in heat production through sympathetic nervous stimulation of the specialised brown fat found in infants in the neck, back, axillae and trunk. This fatty tissue is metabolically very active with the capacity for ample heat production. In spite of this mechanism and a capacity for peripheral vasoconstriction to reduce heat loss comparable to the adult (p. 177) the newborn baby is at risk from hypothermia and needs care and supervision. At about one year the brown fat system is replaced by shivering as the basis of heat production under stress. Reaction to high ambient temperatures is less of a problem in temperate zones and the sweating mechanism for heat loss (pp. 65–6) only matures several months later.

The foetus swallows amniotic fluid from the third month and ingests 500 ml daily at term. This fluid is largely obtained from its own urine and the fluid exchange ensures that normal motility of the stomach and intestine is established. Digestive enzymes are secreted and at delivery the digestion and assimilation of carbohydrates and protein are well established. Bile salts and lipase are deficient and fat absorption is impaired until the age of two or three months. The foetal gut contains green, sticky *meconium* consisting of intestinal secretions, amniotic fluid residue, bile and dead cells. It is evacuated one or two days after delivery.

The newborn baby can suck and swallow effectively with only occasional regurgitation or inhalation of fluid. Digestion is fairly efficient but about 30 per cent of a bottle feed is lost in the stools, largely because of malabsorption of fat. There is rapid improvement over the next few months.

The liver begins to cope with its life-long tasks of synthesis and excretion almost at once. Initially it has difficulty in coping with the increased load of bilirubin resulting from increased red cell breakdown in the newborn. Its ability to conjugate bilirubin and transform it from fat-soluble to water-soluble takes some days or weeks to mature. The fat-soluble form is not excreted as bile and jaundice may result. In severe cases the fat-soluble bilirubin may cross the blood-brain barrier and cause permanent damage to the basal ganglia (p. 143).

The immature liver may not produce sufficient prothrombin and blood clotting factors. This can be partially corrected by injections of vitamin K but serious haemorrhage may occur. This aspect of liver function can take several months to achieve normality.

In foetal life the kidneys are active, producing large amounts of dilute urine with a fairly high sodium content. At birth the kidneys are immature and many of the nephrons are not fully developed. Conservation of sodium with a reduction of sodium excretion is soon established but for several months the infant kidneys cannot deal satisfactorily with an increased load of sodium. This may readily happen if bottle feeding is not scrupulously controlled. In such infants water retention, convulsions and even death may result.

Immediately after birth there is rapid bacterial colonisation of the gastro-intestinal tract, skin and mucous membranes which will persist throughout life. In the breast-fed there is a preponderance of lactobacilli but otherwise the bacterial pattern is determined by the genital tract of the mother and the environmental organisms. The newborn are very liable to dangerous infections because they have little natural immunity and the passive immunity derived from the mother is soon lost. At this stage bacteria which may be harmless to adults can prove dangerous. The infant skin and mucous membranes are delicate and the umbilical stump takes time to heal.

Growth and development

The infant grows steadily but increasingly slowly during the first ten years of life and growth is fastest *in utero* up to the time of delivery. There is a progressive increase in size with developing sensory, motor and intellectual capacities. Development of the various tissues and organs occurs harmoniously but at different rates. The relative proportions of the tissues and organs changes profoundly. The heart and liver grow steadily by cell division throughout childhood but the brain has its full complement of neurones by the age of 5 years. It starts as a flat sheet of cells which folds into a tube and then enlarges steadily at the rate of about 250 000 neurones formed every minute. In adult life there is a steady annual loss of neurones with increasing age. There is no replacement of neurones in later life but liver cell numbers are rapidly restored after damage or loss. The kidney begins to function about ten weeks after conception and at birth has its final total of nephrons, albeit functionally immature.

In the newborn the head is relatively large but in a few years the trunk and legs outgrow the head. The content of body fat is 25 per cent at three months but normally only 10 per cent in the adult. Skeletal muscle increases in amount from 25 to 40 per cent of the adult body. The brain forms over 10 per cent of the mass of the newborn baby but in spite of its huge increase in functional capacity it forms only 2 per cent of the adult body. Around puberty there are profound changes associated with the establishment of the secondary sexual characteristics. The nasal sinuses develop and the jaws and larynx enlarge especially in the male.

Infants and children should be measured annually to reveal any anomalies of growth and to detect any early indications of disease. The supine length is measured up to the age of two years and then the standing height is recorded. Skinfold thickness taken with special forceps is a useful check on nutrition. Testicular volume can be estimated by comparison with standardised ovoid beads. Skeletal maturity can be assessed from an X-ray of one hand and wrist showing the degree of ossification of the epiphyses.

Height depends upon the interaction of several genes with physiological factors such as hormones, sleep and emotional security plus environmental influences like food and exercise. Various infective and intestinal diseases are also vitally important, e.g., tuberculosis, chronic malaria and malabsorption syndromes. There are obvious ethnic variations and in developed countries during the past hundred years there has been an increase in stature and weight with each generation.

The velocity of growth varies during the year and is greatest in spring. The adolescent spurt at puberty occurs 1–2 years earlier in girls, who are thus usually shorter than boys. The spurt in boys is due to androgen secretion and in girls to oestrogens. Growth which threatens to be excessive in girls can be checked by the early induction of puberty by small doses of oestrogens. Growth hormone from the pituitary gland is the essential agent. It is released by various influences and physical exercise, emotional stability and deep sleep are all important. Growth hormone does not act directly but stimulates the liver to produce *somatomedin*, which is the active agent.

Normal babies measure 45–55 cm at term. They increase by 50 per cent at one year and double their length at four years. The birth length increases three times by the age of twelve and to about four times at maturity. There is great individual variation and normal children can be at least 10 per cent above or below the conventional average figures. Mature height can be predicted to some extent by adding the height of the two parents in cm, adding 12 cm for boys or subtracting 12 cm for girls, and dividing the result by two. This method is not very reliable.

There is much greater variation in the acquisition of skills, which may be motor, social, adaptive or linguistic. Delay at one of the so-called 'milestones' of development is often made up later. It is very difficult to define normality because of the continuous interplay between genetic endowment and environment. Speculation about the precise amount of genetic influence in the development of skills and 'intelligence' is therefore futile. It seems that the various skills appear when the central nervous system has matured to that point. Precocity or delay because of environmental factors tend to even out later.

Motor skill development can be followed fairly easily. Mass movements of the foetus appear at about two months, to be followed by sucking and turning at three months. At four months limb withdrawal can be observed and at birth stepping and grasping responses are present. Most infants can sit unsupported between four and eight months; purposive use of finger and thumb appears at about nine months. Walking is achieved anywhere between seven and twenty months.

The neuronal patterns underlying these skills are determined genetically and are mobilised and refined by continuous use, trial, experiment and play. The bonding to the parent which should occur at birth, plus opportunity, encouragement, affection, stability and variety of stimulus are all essential to normal development. It is at this stage that the coordination of skill under the influence of the cerebellum appears and infancy is one long training session.

The mechanism of learning is not clear. It was thought that permanent records were established by the synthesis of new DNA molecules but this seems unlikely as a major factor. More probably learning results from the growth of synaptic contacts between neurones and by the creation of innumerable new synaptic connections as a response to activity. Most of the neurones have been formed by the fourth or fifth month after birth but synaptic proliferation goes on much longer. The cerebellum develops fastest and shows the greatest increase in dendritic branching and synapses. Growth of glia cells and production of myelin come later and brain growth continues up to two years of age.

The development of language

Communication between adults involves gesture, expression, posture and movement as well as speech. After a few weeks a baby begins to respond to movement of objects and individuals, soon responding more strongly to its mother. By six months there is evident vocalisation and a display of emotion, and at one year the mother, at least, can distinguish the first words. The rate of development of speech depends greatly on the social situation and the attentive skill of the mother in responding to the infant. An average infant will use about 200 words at eighteen months and begins to comprehend and combine words at the age of two years. The first sentences appear soon afterwards, and language and comprehension elaborate from then on in response to environmental stimulation.

Puberty

Puberty is the period of achievement of sexual maturity which marks the transformation of the child into the young adult. The process is started by the release of two gonadotropic hormones, FSH and LH, by the anterior pituitary gland in response to hormones from the hypothalamus. The ovary responds to follicle stimulating hormone (FSH) by enlarging and producing ovarian follicles. Oestrogens are secreted and stimulate the growth of the genitalia and the development of secondary sexual characteristics with enlargement of the breasts and the appearance of female body hair. The menstrual cycle is gradually established and luteinising hormone (LH) develops the ovarian follicle into a corpus luteum as described on page 191.

In the male FSH produces enlargement of the seminiferous tubules with development of the germinal cells and finally the full capacity to produce spermatozoa. LH causes proliferation of the interstitial cells which produce testosterone, the male hormone responsible for the appearance of the secondary sexual characteristics. In both sexes the action of the sexual hormones (androgens) is supplemented at puberty by increased secretion of androgens from the suprarenal glands under the influence of adrenocorticotropic hormone from the anterior pituitary.

During this period profound emotional changes occur in parallel with changes in the size and form of the body. Emotional maturation continues for many years but by the late teens physiological processes in both sexes have acquired the characteristics of the adult, although further development in bulk, power and endurance occur.

The ageing processes

During the first thirty years of life the processes of growth, development and repair outstrip the effects of wear, injury and disease. The peak is soon passed and ageing begins and slowly accelerates. The joints become stiffer and the muscles weaker. The exercise capacity and energy reserves slowly decrease. The skin and arteries become less elastic. Recovery from illness or injury becomes slower and less complete. As the body fails, the lessening of physical and mental activity itself leads to further deterioration.

With age the skin sags and wrinkles, the teeth fall out and reproductive powers disappear, entirely in menopausal women and gradually in men. The skeleton becomes brittle, the hair fades and falls, the senses deteriorate and mental powers slowly diminish. Many of these changes result from failing hormonal stimulation. With age the arteries grow narrower and less blood can reach the tissues and organs, and so their metabolic activity decreases. The heart is less able to increase its output on demand. The chest wall becomes more rigid and the lungs less elastic so that the respiratory capacity falls. In many people the slow processes of physical deterioration are terminated by disease or accident. In others degeneration proceeds inexorably, although in widely varying degrees, until death may result from an almost trivial cause.

Several theories have been proposed to explain the phenomena of ageing. Death from old age is rare among wild animals but is becoming increasingly common in humans. The crude death rate fell from about twenty per thousand in 1870 to twelve per thousand in 1930. Improved social conditions of housing, nutrition, water supply and sewage disposal have been mainly responsible for the fall in death rate, but immunisation and antibiotics have probably helped. The greater part of the improvement has been amongst the young. The life expectancy at age 65 is only slightly better today than it was one hundred years ago but because of improved survival among the young there will be about 25 per cent more people over 65 in Britain in ten years' time than now.

It has been claimed that old age is simply due to the wearing out of those parts of the body which cannot be replaced. For example, the kidney at age 70 has lost more than one third of its nephrons. There is also a progressive deterioration in the blood supply to all the organs and tissues due to silting up of the arterial system by the ubiquitous atheroma. In addition there is a steady deposition of metabolites in the cells and tissues shown by collections of pigment and abnormal collagen fibres. It is probable, also, that the production of hormones by the endocrine glands becomes diminished and unbalanced, undermining the essential homeostatic mechanisms already described. Finally it is likely that the lifelong process of DNA replication will show a gradually increasing number of errors of transcription because of mutation or intracellular damage during protein synthesis. This would eventually lead to the accumulation of abnormal proteins and to increasing deficiencies in the defensive immunological systems.

Whatever the causes, death is finally inevitable and socially beneficial.

Index